T0199711

Mathematical Techniques
for Wave Interaction with
Flexible Structures

IIT Kharagpur Research Monograph Series

Published Titles:

Modeling of Responsive Supply Chain, *M.K. Tiwari, B. Mahanty, S. P. Sarmah, and M. Jenamani*

Micellar Enhanced Ultrafiltration: Fundamentals & Applications, *Sirshendu De and Sourav Mondal*

Microfluidics and Microscale Transport Processes, *edited by Suman Chakraborty*

Mathematical Techniques for Wave Interaction with Flexible Structures, *Trilochan Sahoo*

IIT KHARAGPUR RESEARCH MONOGRAPH SERIES

Mathematical Techniques for Wave Interaction with Flexible Structures

Trilochan Sahoo

CRC Press
Taylor & Francis Group
Boca Raton London New York

CRC Press is an imprint of the
Taylor & Francis Group, an **informa** business

CRC Press
Taylor & Francis Group
6000 Broken Sound Parkway NW, Suite 300
Boca Raton, FL 33487-2742

First issued in paperback 2019

ISBN-13: 978-1-4665-0604-6 (hbk)
ISBN-13: 978-0-367-38075-5 (pbk)

Library of Congress Cataloging-in-Publication Data

Sahoo, Trilochan.
   Mathematical techniques for wave interaction with flexible structures / Trilochan Sahoo.
       p. cm. -- (IIT Kharagpur research monograph series)
   Includes bibliographical references and index.
   ISBN 978-1-4665-0604-6 (hardback)
   1. Fluid-structure interaction--Mathematical models. 2. Flexible structures--Mathematical models. I. Title.

TA357.5.F58S26 2012
624.1'71--dc23   2012028458

Visit the Taylor & Francis Web site at
http://www.taylorandfrancis.com

and the CRC Press Web site at
http://www.crcpress.com

Dedicated to my parents
Sri Lingaraj Sahoo and Srimati Hara Sahoo

Contents

List of Figures

About the Series

IIT Kharagpur had been a forerunner in research publications and this monograph series is a natural culmination. Empowered with vast experience of over sixty years, the faculty now gets together with their glorious alumni to present bibles of information under the *IIT Kharagpur Research Monograph Series*.

Initiated during the Diamond Jubilee Year of the Institute, the Series aims at collating research and development in various branches of science and engineering in a coherent manner. The Series, which will be an ongoing endeavour, is expected to function as a source reference to fundamental research as well as to provide directions to young researchers. The presentations are in a format that can serve as stand alone texts or reference books.

The specific objective of this research monograph series is to encourage the eminent faculty and coveted alumni to spread and share knowledge and information to the global community for the betterment of mankind.

The Institute

Indian Institute of Technology Kharagpur is one of the pioneering Technological Institutes in India and it is the first of its kind to be established immediately after the independence of India. It was founded on 18 August, 1951, at Hijli, Kharagpur, West Bengal, India. The IIT Kharagpur has the largest campus of all IITs, with an area of 2,100 acres. At present, it has 34 departments, centers and schools, and about 10,000 undergraduate, postgraduate and research students with faculty strength of nearly 600; the number of faculty is expected to double within approximately five years. The faculty and the alumni of IIT Kharagpur are having wide global exposure with the advances of science and engineering. The experience and the contributions of the faculty, students and the alumni are expected to get much exposure through this monograph series.

More on IIT Kharagpur is available at www.iitkgp.ac.in

Preface

The area of wave structure interaction is a major branch of fluid mechanics and assumes special significance due to its wide range of application in Marine Technology and Arctic Engineering. The objective of the proposed monograph is to present a contemporary account of various mathematical techniques on wave structure interaction problems and to deal with a class of initial/boundary value problems associated with Laplace equation satisfying certain higher order boundary conditions. These classes of boundary value problems are not of the Sturm-Liouville type in nature and are rarely discussed in any text books. All efforts are made to present the material in the monograph in a pedagogical manner to enable the readers to appreciate various tools discussed and apply them to their own problems in related branches. At the end of each chapter, certain exercises are left for the readers; some of these exercises are not yet attempted for solution. The content is divided into seven chapters based on the application of the various mathematical techniques and the discussed physical problems.

Chapter 1 deals with general introductions and is concerned with the motivation, the state of the art, a brief introduction to differential equations, and fundamentals of wave structure interaction problems. The objective of this chapter is to help those unfamiliar with water waves and flexible structures but who are interested in the various mathematical tools with a certain understanding of the physical problems. A major part of the materials presented here are available in different research papers and text books in a scattered manner. Chapter 2 is concerned with the application of Fourier analysis to deal with boundary value problems associated with Laplace equations satisfying higher order boundary conditions. General expansion formulae are discussed for wave structure interaction problems and various characteristics of the associated eigenfunctions are briefly described. In Chapter 3, Green's function techniques are used to derive alternate derivation of the expansion formulae. Further, Green's function and source potentials associated with wave structure interaction problems are derived in several cases. Chapter 4 is concerned with the role of wave interaction with vertical flexible structures in the cases of both single- and double-layer fluids having a free surface. Eigenfunction expansion methods along with least square approximations are used to deal with the physical problems. Applications of the wide spacing approximation method for wave interaction with multiple structures are illustrated. In Chapter 5, mathematical approaches to deal with initial boundary value problems associated with surface wave interactions with large floating structures are discussed in brief. Applications of combined Laplace, Fourier and Hankel transforms along with the method of stationary phase in solving transient problems are discussed and illustrated for different cases. Chapter 6 deals with wave structure interaction problems in case of long wave under the assumption of linear shallow water approximation. Apart from eigenfunction expansion methods, wide spacing approximations are used to solve various problems. In Chapter 7, the use of the boundary integral equation method is illustrated by analysing the wave diffraction by a flexible floating membrane. In this chapter, the boundary value problems are converted into integral and/or integro-differential equations with the help of Green's function, and their solutions are obtained by approximate methods via two types of approximations.

Although emphasis is on physical problems associated with water waves, the underlying mathematical tools can be easily extended to deal with physical problems in the areas of acoustics, electromagnetic waves, wave propagation in elastic media and solid-state physics. Research scholars and graduate students of applied mathematics, computational/theoretical physics, wave structure interaction, hydroelasticity, ice physics and ice engineering will find the monograph useful. The monograph will act as a handbook for naval architects and ocean engineers dealing with the design of floating and/or flexible marine structures and geophysicists working on wave ice interaction problems. The monograph can be used as a textbook related to a course on wave structure interaction or hydroelasticity at the postgraduate level.

Thanks are due to colleagues in the Department of Ocean Engineering and Naval Architecture and authorities of IIT, Kharagpur for proposing to write this monograph during the Diamond Jubilee year of the Institute. Special thanks are due to the late Professor A. T. Chwang of the Department of Mechanical Engineering, The University of Hong Kong, whose support and encouragement inspired the author in many ways to pursue this branch of study. The author thanks Ramnarayan, Sanjay, Sourav and Harekrushna for their help in checking some of the calculations. Specially, thanks are due to Mr. Sarat Chandra Mohapatra, who has helped in preparing various figures and checking various parts of the monograph. The author wishes to thank the CRC Press for agreeing to bring out this monograph. In particular, thanks are due to the project coordinator Amber Donley, the project editor Michele Dimont, and Dr. Gagandeep Singh of CRC Press for their encouragement and support during the preparation of the manuscript.

The author gratefully acknowledges his brothers Bhagaban and Chaitanya and elder sister Baidehi for their forbearance, support and encouragement. Above all, special thanks and love are due to my wife Sagarika, son Om and daughter Shrivali for their unconditional love and support. It was a long and difficult journey for them during the preparation of the manuscript.

Trilochan Sahoo

1

General introduction

1.1 Preamble

Wave structure interaction problems involve the understanding of the fluid phenomena which is in contact with the atmosphere and the behavior of the structure that is in contact with the fluid in addition to the phenomenon of wave propagation. This interdisciplinary branch embodies the basic equations of fluid mechanics, structural mechanics, concepts on wave propagation and the critically important role of some special types of boundary conditions. There is a parallel interest in analysing wave interaction with a large floating ice sheet which is also modeled as a large floating elastic plate and is of significant importance in the marginal ice zone of the polar region. In many situations, the structures are considered flexible in nature and hydroelastic analysis is made to analyse several hydrodynamic characteristics. Some of these structures have the advantages over the traditional rigid structure because they are reusable, environmental friendly and cost-effective. As a result, a substantial growth of knowledge in the dynamics of ocean surface waves and their effects on ocean structures has been witnessed in the past two decades.

In the present monograph, the emphasis is on the unique determination of the solution for a class of physical problems associated with the Laplace equation satisfying higher order boundary conditions arising in the broad field of fluid structure interaction. Discussed are the methods based on the application of the theory of ordinary and partial differential equations, Fourier analyses, complex function theory, Green's function techniques, the boundary integral equation method, the least square approximation method, the wide spacing approximation method, integral transform methods and the method of stationary phase. A basic mathematical background of ordinary differential equations, partial differential equations, complex function theory and numerical analysis at the graduate level will be of immense support in understanding various mathematical tools discussed in the monograph. A basic understanding of the equation of fluids and structures will be helpful to appreciate the physical problems discussed in the monograph. However, a section in Chapter 1 is earmarked to enable the readers to be familiar with the basic equations of fluids and structures and the basics of water wave motion. All the methods are illustrated by considering physical problems associated with gravity wave interaction with floating and/or flexible structures arising in ocean engineering and arctic engineering/geophysics. Although, the subjects of finite amplitude ocean waves and extreme wave climates are of significant importance, emphasis in this monograph is almost exclusively limited to small amplitude surface waves and structural response for linear analysis to be applicable. The structures of our interest are mainly flexible strings/membranes and elastic beams/plates in the present monograph. The Euler-Bernoulli beam equation or the Timoshenko-Mindlin equation, as appropriate to physical problems, will be used in the modeling of the elastic plate.

1.2   State of the art

The wave structure interaction problems discussed in the present monograph fall in a major branch of the subject hydroelasticity. The term *hydroelasticity* first appeared in the technical literature of the first symposium of Naval Structural Mechanics sponsored by the Office of the Naval Research and Stanford University in 1958, and is analogous to aeroelasticity [47]. Although, the subject of hydroelasticity started with keeping floating vessels in mind, the related principles and methodologies are now being applied to a wide variety of marine structures [13]. However, in the recent decades, the theory of wave structure interaction finds its application in various areas of marine technology and arctic engineering such as motion analysis of high-speed vessels; very large floating structures (VLFSs) and mobile offshore bases (MOBs) for efficient utilization of ocean space for humanitarian and military operations such as floating airports, floating bridges and buoyant tunnels; marine risers; cable systems and umbilicals for remotely operated or tethered underwater vehicles; seismic cable systems; flexible containers for water transport, oil spill recovery and other purposes; floating and submerged flexible breakwaters; interaction of sea waves with ice fields; silt curtain used for creating tranquility zones, oil spill recovery and many other uses in ports and harbors; and mooring lines. The main advantages of the wave interaction with flexible structures are the accurate idealization of the fluid structure interaction system and more rigorous analysis by which dynamic responses, such as the stresses and bending moments, in waves are obtained.

In recent decades, the subject of wave structure interactions has gained considerable importance to analyse wave interaction with VLFSs and mobile offshore bases (MOB) for utilization of ocean space for various humanitarian and military operations. Constructions of these structures are cost-effective when the water depth is great as they are easy and fast to construct and can be removed or expanded with ease. The VLFSs are protected from the seismic shocks as the energy can be dissipated into the infinite ocean. These structures do not damage the marine ecosystem, fill up the deep harbors with sediments or obstruct the ocean current and thus are environmentally friendly. There are several reviews in the literature on significant progress on wave interaction with flexible floating structures in homogeneous fluid and structural medium in cases of uniform water depth (see [67], [141] and [16]). Various aspects of wave interaction with large floating elastic plates have been studied in [106], [130], [49] and the literature cited therein. A comparison of various computer codes based on several fluid and structural models for hydroelastic analysis of ships and VLFSs is given in [118]. The optimum location and rotational stiffness of the connectors for the two–floating beam system with the view to minimise the hydroelastic response is studied in [119]. It may be noted that the global response of an interconnected floating structure can be effectively reduced by adjusting the connector properties. There is a recent review on the developments of various procedures to mitigate the hydroelastic response of VLFS under wave action (see [139]). In addition, a review on the developments and achievements of linear and nonlinear 3D hydroelasticity theories of ships, and the associated numerical and experimental techniques are given in [149]. Also, a review on the recent advances and future trends in hydroelasticity of ships can be found in [52]. Recently, a review on various modeling challenges on hydroelasticity was covered in [71].

The problems of wave–ice interactions play an important role in the Arctic and Antarctic regions, where large floating ice fields interact with the ocean waves. The floating ice sheets are modeled as flexible structures that deform under the ocean wave loading. This will lead to more destructive wave loads on the less compact fields of packed ice that have already been weakened by elevated temperatures. The study on wave interaction with floating ice sheets started in 1887 ([42]). These floating ice sheets are also modeled as thin floating elastic plates and are assumed to be very large floating flexible structures. The thickness of this kind of structure is negligible (less than 10m) compared to its length (on the order of kilometers), and the thin plate model of Euler-Bernoulli is in general used in the mathematical formulation of the boundary value problem. As the gravity waves from the open ocean penetrate into the ice shelf, they experience an impedance change due to the flexural properties of the structure in addition to several irregularities in the medium such as changes in the ice thickness or in physical and mechanical properties of the ice. These irregularities result in altering the dispersion relation associated with the wave propagating below the ice sheet and hence a transformation of the waves takes place below the ice sheet [147]. Because of the commonality of the two classes of problems, the study on wave interaction with large floating structures plays a vital role in both marine technology and arctic engineering. Several aspects of wave–ice interaction are reviewed in [124]. Commonality of wave–ice interaction and wave interaction with very large floating structures are discussed in [125].

Recent emphasis on flexural gravity waves deals with problems having heterogeneous boundaries. The surface wave diffraction by a crack in an ice sheet with the additional assumption of abrupt change in the ice thickness across the crack is solved in [91] based on the Wiener-Hopf technique. Flexural gravity wave propagation at the interface between two floating ice sheets of different flexural rigidity by the application of Wiener-Hopf technique is investigated in [21]. A historical perspective of the Wiener-Hopf technique can be found in [78]. A solution for the linear wave forcing of a floating thin elastic plate on water of variable depth is derived in [138] by the combined application of boundary element method and the integral equation method. Influence of water depth on the hydroelastic response of VLFS is studied in [2], [9], [61] and the literature cited therein. Various aspect of the wave structure interaction problems are studied in the literature (see [6], [8], [46], [60], [113], [146] for further details).

The mathematical problems associated with the class of physical problems discussed above give rise to a class of boundary value problems associated with the Laplace equation satisfying higher order conditions on the structural boundaries. Thus, the associated eigenfunctions are not orthogonal in the usual sense, and the boundary value problems are not of the Sturm-Liouville type in nature. Similar problems arise in wave interaction with floating ice sheets in the polar region (as in [126] and the literature cited therein). On the other hand, in acoustic structure interaction problems (see [75], [76] and cited literature), higher order boundary conditions occur on the structural boundary for a class of problems associated with the Helmholtz equation.

Recently, expansion formulae and the associated orthogonal relations have been derived for a large class of boundary value problems associated with the Laplace/Helmholtz equation satisfying higher order boundary conditions arising in the area of acoustics. These orthogonal relations are suitable for a large class of boundary value problems in infinite/semi-infinite strip type domains. Significant progress has been made in developing expansion formulae to deal with wave structure interaction problems (see [77], [121], [87], [59], [9], [101]). Some of the most significant properties of the eigenfunctions for a class of problems associated with the Helmholtz equation satisfying higher order boundary conditions arising in wave guide problems in the field of acoustics are given in [74] and can be easily extended to a class of problems associated with the Laplace equation satisfying higher order boundary conditions. Further, expansion formulae to deal with wave structure interaction problems in three dimensions satisfying higher order boundary conditions on the structural boundaries associated with the Helmholtz equation are derived by Lawrie ([75]). Recent developments on the expansion formulae are reviewed in [76]. Further, expansion formulae for the velocity potential for the 3D Laplace equation satisfying higher order boundary conditions associated with wave structure interaction problems are derived in several cases in semi-infinite strips and quarter planes in [103]. Recently, a review on various modeling challenges on hydroelasticity was published by [71].

To deal with wave structure interaction in a two-layer fluid having free surface and interface, expansion formulae were developed in [88] which are generalised in [100] to deal with wave motion in the presence of surface and interfacial tension. Expansion formulae associated with flexural gravity waves in the presence of an interface are derived in [10] and [104]. Further generalisations on expansion formulae can be found in [76].

There are two major approaches for dealing with the hydroelastic analysis of very large floating structures, namely frequency domain analysis and time domain analysis. In the case of pontoon-type VLFSs, the wave motion and the structural responses are assumed to be small, and in these cases, the analysis is based on frequency domain analysis. On the other hand, for a transient response including nonlinear motion due to waves, it is essential to perform the time domain analysis found in [67] and [141]. However, in the present monograph, our major concern is on the frequency domain analysis in which the physical problem reduces to a boundary value problem associated with the Laplace equation satisfying certain higher order boundary conditions. There has been significant progress on time domain problems in the last decade by suitable application of integral transform methods and methods of steepest descent, the fast Fourier transform method, the spectral method and the suitable mode expansion method ([135] and the cited references).

In the theoretical analysis of large floating structures and ice sheets, the structure is modeled as a large floating elastic plate of negligible thickness as the thickness of the structure is very small compared to the length of the structure. Thus, the depth of submergence of the structure is neglected in the mathematical modeling in most of these cases. Due to the largeness of the structure, the computational burden becomes too cumbersome so that it is often difficult to carry out the analysis. Thus, to overcome these difficulties often these types of structures are assumed to be infinitely long in comparison with the wavelength of the incident waves. In the analysis of a wide class of problems, marine bodies are in general assumed to be rigid. However, the elastic effects of the structures are to be taken into account under certain wave conditions when (i) the body itself is flexible, (ii) the body is very thin compared to wave parameters and (iii) the body is very long with respect to the incident waves. Although, the first two conditions are very obvious, in the last case, the localized deflection or vibration of a long structure becomes significant due to the continuous excitation of small amplitude waves, although the motion of the whole body is small as compared to its length (see [121]). Apart from the use of submerged horizontal structures for attenuation of wave height in open sea in the marine environment (as in [43]), submerged horizontal structures are proposed for attenuation of structural responses of large floating structures (as in [139] and the literature cited therein).

In contrast to the utilization of space and application in cold region science and technology, there is a wide interest in the use of floating and submerged flexible structures for the purpose of wave attenuation, oil spill recovery and other applications in coastal engineering in deep and shallow seas. These structures have the advantages over the traditional rigid structure in that they are reusable, environmental friendly and cost-effective. The elastic structure is modeled as a vertical flexible plate. However, both the horizontal and vertical flexible membranes are used for coastal engineering application purposes (see [73], [136], [152] for further details).

Another important branch of study in the field of ocean engineering is the interaction of progressive waves with ocean currents, which increase wave height and steepness, thus adding to the hazards of navigation and damage to ships ([57]). Currents in the ocean are generated due to forces acting upon water such as the earth's rotation, change in wind direction, temperature and salinity differences and gravitation of the sun or moon. Currents with speeds exceeding 1m/s are observed in most of the coastal areas of the world. Due to the mutual interaction between waves and the underlying currents, the wave characteristics such as wave length, wave period and hence the phase and group velocities change significantly and that leads to a change in the structural load for all types of marine structures. An extensive study on wave current interaction can be found in [112]. In addition, there has been significant progress on wave interaction with flexible structures in the presence of current (as in [123], [27], [98], [97], [86]). The effect of uniform current on the wave motion in the presence of a floating ice sheet has been studied recently (see [9], [11], [12]).

Various mathematical methods for a broad class of engineering problems are discussed in [94]. Various mathematical aspects on wave structure interaction (discussed in [82]) are meant for rigid structures and the Sturm-Liouville type of boundary value problems. In the recent book [90], Chapter 9 deals with the scattering of surface waves by flexible structures based on the application of integral equations. Further, in Chapter 6 of [51], wave interactions with floating flexible platforms are discussed. However, in this book, all of the chapters will be dealing with mathematical approaches for problems of wave interaction with flexible structures associated with the non-Sturm-Liouville type of boundary value problems.

1.3   Fundamentals of differential equations

In this section, some of the basic definitions and results related to initial and boundary value problems associated with ordinary and partial differential equations are introduced. Further details can be found in various text books on differential equations.

1.3.1   Introduction to ordinary differential equations

There is a large class of physical problems which can be easily reduced to ordinary differential equations. Thus, the theory and solution techniques for ordinary differential equations play a significant role in solving a wide class of problems of mathematical physics and engineering as well as problems of biology and social science.

Definition 1 A linear differential equation (DE) of n-th order is of the form

$$L[y] \equiv (a_n D^n + a_{n-1} D^{n-1} + ... + a_1 D + a_0)y = f(x), \quad a < x < b \quad (1.1)$$

where a_i for $i = 0, 1, 2, ..., n$ are known constants and D^i denotes the i-th order differential operator and is given by

$$D^i = \frac{d^i}{dx^i}$$

for $i = 1, 2, ..., n$. The operator L is called a linear differential operator and the differential equation is called a linear differential equation with constant coefficients. Further, if $f(x) = 0$, then the DE is called homogeneous; otherwise it is called nonhomogeneous.

Lemma 1 The linear differential operator L satisfies the superposition principle given by

$$L\left[\sum_{i=1}^{n} c_i y_i\right] = \sum_{i=1}^{n} c_i L[y_i], \quad (1.2)$$

with c_i being known constants.

Definition 2 If $y = e^{mx}$ is a solution of the homogeneous DE given by

$$L[y] \equiv (a_n D^n + a_{n-1} D^{n-1} + ... + a_1 D + a_0)y = 0, \quad (1.3)$$

then the algebraic equation

$$a_n m^n + a_{n-1} m^{n-1} + ... + a_1 m + a_0 = 0 \quad (1.4)$$

is called the characteristic equation of the DE, which will have n roots. Depending on the nature of the constants a_n, the algebraic equation in Eq. (1.4) will have real and distinct, real and repeated or complex roots. Next, without any proof, we will mention the general solution in different cases.

Lemma 2 If the roots of the characteristic equation (1.4) are real and distinct and are given by $m_1, m_2, ..., m_n$, then $e^{m_1 x}, e^{m_2 x}, ..., e^{m_n x}$ are n distinct solutions of the DE (1.3) and these solutions are linearly independent in the interval [a, b]. The general solution of the DE (1.1) is given by

$$y(x) = c_1 e^{m_1 x} + c_2 e^{m_2 x} + ... + c_n e^{m_n x}, \quad (1.5)$$

where $c_1, c_2, ..., c_n$ are arbitrary constants.

Lemma 3 If the roots of the characteristic equation (1.4) are real and one of the roots m_1 appears r times, $m_{r+1}, m_{r+2}, ..., m_n$ are the other $n-r$ roots being real and distinct, then the general solution of the DE (1.3) is given by

$$y(x) = (c_1 + c_2 x + c_3 x^2 + ... + c_r x^{r-1})e^{m_1 x} + c_{r+1}e^{m_{r+1} x} + ... + c_n e^{m_n x}, \quad (1.6)$$

where $c_1, c_2, ..., c_n$ are arbitrary constants.

Lemma 4 If the characteristic equation (1.4) has a complex root of the form $\alpha + i\beta$, then its complex conjugate $\alpha - i\beta$ will also be a root of the characteristic equation and the corresponding part of the general solution is given by

$$y(x) = e^{\alpha x}(c_1 \cos \beta x + c_2 \sin \beta x). \quad (1.7)$$

Corollary 1 If the two roots m_1 and m_2 of the characteristic equation (1.4) are given by $\alpha \pm i\beta$, $m_3 = m_4$ and the other roots are real, distinct and different, then the general solution is given by

$$y(x) = e^{\alpha x}(c_1 \cos \beta x + c_2 \sin \beta x) + (c_3 + c_4 x)e^{m_3 x} + c_5 e^{m_5 x} + ... + c_n e^{m_n x} \quad (1.8)$$

where $c_1, c_2, ..., c_n$ are arbitrary constants.

Theorem 1 *The full solution of the nonhomogeneous DE (1.1) is the sum of the particular solution y_p of the nonhomogeneous DE and the general solution y_{CF} of the homogeneous DE (1.3). The solution y_{CF} is known as the complementary function and the solution y_p is called the particular integral.*

The differential equation (1.1) can be rewritten as

$$F(D)y = f(x), \quad (1.9)$$

where $F(D) = a_n D^n + a_{n-1}D^{n-1} + ... + a_1 D + a_0$. Few results in terms of lemmas and theorems concerning the particular integrals are highlighted here which are used frequently in the monograph.

Theorem 2 *Consider the $(2l_0 + 1)$-th order DE given by*

$$\{\mathcal{L} + \mathcal{M}\}y = \sin kx, \quad k > 0, \ a < x < b, \quad (1.10)$$

where $\mathcal{L}$ and $\mathcal{M}$ are linear differential operators of the form

$$\mathcal{L} \equiv \sum_{j=0}^{l_0}(-1)^j a_j D^{2j+1}, \quad \mathcal{M} \equiv \sum_{j=0}^{m_0}(-1)^j b_j D^{2j}, \quad (1.11)$$

with a_j and b_j being known constants, l_0 and m_0 are positive integers with $m_0 \leq l_0$; $P(k; l_0) = \sum_{j=0}^{l_0}(-1)^j a_j k^{2j+1}$ and $Q(k; m_0) = \sum_{j=0}^{m_0}(-1)^j b_j k^{2j}$ are the characteristic polynomials associated with the differential operators $\mathcal{L}$ and $\mathcal{M}$, respectively. The particular integral of the DE in Eq. (1.10) is given by

$$y_p = -\frac{P(k;l_0)\cos kx - Q(k;m_0)\sin kx}{P^2(k;l_0) + Q^2(k;m_0)}, \quad k > 0,\ a < x < b, \quad (1.12)$$

where $P^2(k;l_0) + Q^2(k;m_0) \neq 0$.

From Theorem 1, it is obvious that the general solution involves n arbitrary constants. Thus, the unique determination of the solution requires n initial/boundary conditions to be prescribed depending on the nature of the problem at hand.

Definition 3 The problem of finding a solution $y(x)$ of the DE (1.1) in a closed interval $[a, b]$ satisfying the initial conditions

$$y(x_0) = y_0, y'(x_0) = y_1, ..., y^{n-1}(x_0) = y_{n-1}$$

is called an initial value problem. The conditions at $x = x_0$ are called the initial conditions with x_0 being a known constant, and the data $y_0, y_1, ..., y_{n-1}$ are assumed to be known. From theory of DE, it is known that a unique solution of an n-th order linear DE is obtained by prescribing n conditions at one point (for an initial value problem).

Definition 4 The problem of determining the solution of the n-th order differential equation (1.1) satisfying n boundary conditions is called a boundary value problem. In this case, the dependent variable satisfies subsidiary conditions at n points. Unlike in the case of initial value problems, in this case, the data may be given in the domain of consideration including the end points. In case of a boundary value problem, it may be possible that the boundary value problem has no solution or many solutions.

Definition 5 The boundary value problem in which the boundary conditions are given at only the two end points is called a two-point boundary value problem and is given by

$$y''(x) = f(x, y, y'), \quad a \leq x \leq b, \quad (1.13)$$

satisfying the boundary conditions

$$a_0 y(a) + a_1 y'(a) = \alpha, \quad b_0 y(b) + b_1 y'(b) = \beta, \quad (1.14)$$

where a, b, α and β are known constants.

A large class of physical problems arises in various branches of mathematical physics typically from problems of continuum mechanics which give rise to boundary value problems. These problems represents a class of linear boundary value problems and are in general time-dependent. Assuming that the motion is simple harmonic in nature, these problems on simplification lead to the Sturm-Liouville type of boundary value problems. The mathematical theory of these types of boundary value problems is an important building block in the derivation of a solution of a large class of physical problems associated with boundary value problems of a general nature.

Definition 6 A Sturm-Liouville equation is a second-order linear homogeneous DE of the form

$$(Lu)(x) + \lambda r(x)u(x) = 0, \tag{1.15}$$

where the differential operator $(Lu)(x)$ given by

$$(Lu)(x) = -(p(x)u'(x))' + q(x)u(x). \tag{1.16}$$

Definition 7 A regular Sturm-Liouville system is a Sturm-Liouville equation as in Eq. (1.15) defined in a finite closed interval $a \leq x \leq b$, satisfying the end point conditions of the forms

$$\alpha u(a) + \alpha' u'(a) = 0, \quad \beta u(b) + \beta' u'(b) = 0, \tag{1.17}$$

where $\alpha, \alpha', \beta, \beta'$ are known real nonzero constants. Often, the Sturm-Liouville systems are referred to as Sturm-Liouville boundary value problems. A nontrivial solution of an S-L system is called an eigenfunction, and the corresponding λ is called its eigenvalue. The set of all eigenfunctions of a regular Sturm-Liouville system is called the spectrum of the system.

In the analysis of physical vibration problems, these eigefunctions are associated with the individual modes of vibration/oscillation of various physical problems. The solution techniques for Sturm-Liouville equations are discussed in several textbooks on differential equations and details are differed here. However, one of the important characteristics of Sturm-Liouville type boundary value problems is highlighted in terms of a theorem as below.

Theorem 3 *Let λ and μ be two distinct eigenvalues and u and v be the corresponding eigenfunctions associated with a regular S-L system defined by Eq. (1.15) and (1.17) over the interval $a \leq x \leq b$. Then,*

$$\int_a^b r(x)u(x)v(x)dx = 0, \tag{1.18}$$

where $r(x)$ is the weight function associated with the eigenfunctions. Often the relation in Eq. (1.18) is called the orthogonal property of the eigenfunctions associated with an S-L system.

1.3.2 Introduction to partial differential equations

The theory of the partial differential equation is vast and its impact is felt in almost every branch of scientific endeavor. In this subsection, basic partial differential equations relevant to wave structure interaction problems will be introduced in brief. One of the most important of all partial differential equations is

$$\frac{\partial^2 u}{\partial x^2} + \frac{\partial^2 u}{\partial y^2} + \frac{\partial^2 u}{\partial z^2} = 0, \tag{1.19}$$

known as the Laplace equation. The elliptic partial differential equation

$$\frac{\partial^2 u}{\partial x^2} + \frac{\partial^2 u}{\partial y^2} + \frac{\partial^2 u}{\partial z^2} = f(x, y, z), \tag{1.20}$$

is the Poisson equation. There are many areas of mathematical physics and engineering such as fluid flows, elastic membranes, acoustics, electrostatics, and steady state heat conduction problems, which lead to the solution of the Laplace/Poisson equation in two/three dimensions. The appropriate choice of boundary conditions is of paramount importance in the selection of appropriate mathematical tools to solve the problem of interest. Often the three-dimensional Laplace equation is further simplified by assuming $u(x, y, z) = Re(\phi(x, y)e^{\pm ipz})$, which reduces the three-dimensional Laplace equation to a two-dimensional elliptic partial differential equation often called the reduced wave equation of the form

$$\frac{\partial^2 \phi}{\partial x^2} + \frac{\partial^2 \phi}{\partial y^2} - k^2 \phi = 0. \tag{1.21}$$

On the other hand, the elliptic partial differential equation

$$\frac{\partial^2 u}{\partial x^2} + \frac{\partial^2 u}{\partial y^2} + \frac{\partial^2 u}{\partial z^2} + k^2 u = 0 \tag{1.22}$$

is the well-known Helmholtz equation. This equation is derived from the wave equation

$$\frac{\partial^2 u}{\partial x^2} + \frac{\partial^2 u}{\partial y^2} + \frac{\partial^2 u}{\partial z^2} = \frac{1}{c^2}\frac{\partial^2 u}{\partial t^2} \tag{1.23}$$

by assuming that $u(x, y, z, t) = Re\{u(x, y, z)e^{\pm i\omega t}\}$. In Eq. (1.23), $c = \omega/k$ is known as the phase velocity or the speed of propagation of the wave. The Helmholtz equation arises in wave propagation problems in various branches of physics such as acoustics, electromagnetic theory, elasticity and water waves. Helmholtz equation, being elliptic in nature, is treated in the same manner as the Laplace equation. However, every unique solution of the Helmholtz equation in an unbounded domain has two parts. One is called the regular solution which exists everywhere, and the other solution satisfies the radiation condition given by

$$\lim_{r \to \infty} r^{(n-1)/2}\left(\frac{\partial u}{\partial r} \pm iku\right) = 0, \tag{1.24}$$

where n refers to the dimension of the Hemholtz equation, $i = \sqrt{-1}$ and $r = \sqrt{x_1^2 + x_2^2 + ... + x_n^2}$ with $(x_1, x_2, ..., x_n)$ being any points in the n-dimensional space. The condition with the positive sign corresponds to the outgoing waves and the condition with the negative sign corresponds to the incoming waves. For the three-dimensional Helmholtz equation as in Eq. (1.22), the radiation condition reduces to

$$\lim_{r \to \infty} r \left(\frac{\partial u}{\partial r} \pm iku \right) = 0, \qquad (1.25)$$

where $r = \sqrt{x^2 + y^2 + z^2}$. For determining unique solutions of certain boundary value problems associated with the Helmholtz equation in the unbounded domain, the radiation condition of the type in Eq. (1.24) was initially introduced by Sommerfeld in 1912 [122] and is referred as the Sommerfeld radiation condition in the literature. Irrespective of the bounded/unbounded nature of the physical domain, the general solution of elliptic partial differential equations depends on one or more of the following three types of boundary conditions as discussed below.

Let $u(x, y, z)$ satisfy one of the elliptic partial differential equations in the domain Ω and let $\partial\Omega$ be the boundary of Ω. If the functional values are prescribed on the boundary $\partial\Omega$, the boundary condition is of the form

$$u(x) = f(x), \quad x \in \partial\Omega, \qquad (1.26)$$

where $f(x)$ is assumed to be known. Such boundary conditions are known as Dirichlet-type boundary conditions. On the other hand, if the normal derivatives of the function are prescribed on the boundary, then such conditions are called Neumann conditions. In such a case,

$$\frac{\partial u}{\partial n} = g(x), \quad x \in \partial\Omega, \qquad (1.27)$$

with $g(x)$ being a known function. There are physical problems, in which the boundary conditions are often a linear combination of both Dirichlet- and Neuman-type boundary conditions which are known as Robins conditions. In such a case,

$$\frac{\partial u}{\partial n} + Ku = h(x), \quad x \in \partial\Omega, \qquad (1.28)$$

where K is a known constant and $h(x)$ is a known function. Unlike a particular type of condition being given on the boundary, often on a part of the boundary, the condition is of the Dirichlet type and on the other half the condition is of the Neumann/Robins type. Such types of problems are called mixed boundary value problems. Mixed boundary value problems are very common in wave structure interaction problems.

Another important elementary partial differential equation is the wave equation. Unlike the Laplace equation, in the case of the wave equation, one looks for a real valued function $u(x, y, z, t)$ that depends on the spatial variables and the time variables such that

$$\frac{\partial^2 u}{\partial x^2} + \frac{\partial^2 u}{\partial y^2} + \frac{\partial^2 u}{\partial z^2} = \frac{1}{c^2}\frac{\partial^2 u}{\partial t^2}. \tag{1.29}$$

In this case, the Laplacian acts on the spatial variables and c represents the speed of propagation which depends on the material of the medium. Many physical problems associated with electromagnetic waves, elastic waves and water waves are described through this equation. Since the wave equation is second order in time, usually two initial conditions are prescribed and are given by

$$u(x, 0) = f(x), \quad u_t(x, 0) = g(x). \tag{1.30}$$

Here, also, Dirichlet and/or Neumann conditions are prescribed on the boundary in the space variables. In addition, initial conditions are required to be specified on the time variable t associated with the wave equation. One of the famous initial value problems associated with the one-dimensional wave equation

$$\frac{\partial^2 u}{\partial x^2} = \frac{1}{c^2}\frac{\partial^2 u}{\partial t^2} \tag{1.31}$$

defined as the problem of determining the real valued function $u(x, t)$ defined in the half plane $(x, t) \in (-\infty, \infty) \times (0, \infty)$ subject to the initial conditions

$$u(x, 0) = f(x), \quad u_t(x, 0) = g(x), \tag{1.32}$$

is the well-known Cauchy problem. The solution for the Cauchy problem discussed is given by

$$u(x, t) = \frac{1}{2}[f(x - ct) + f(x + ct)] + \frac{1}{2}\int_{x-ct}^{x+ct} g(s)ds \tag{1.33}$$

and is known as the D'Alembert solution for Cauchy problem. For classical methods such as integral transforms and the eigenfunction expansion method, Green's function techniques are applied to solve a large class of boundary value problems associated with both ordinary and partial differential equations. Apart from these classical methods, methods such as the Wiener-Hopf technique, the method of steepest descent and complex variable methods, numerical methods based on variational principle, boundary element methods and finite element methods are used to deal with initial and boundary value problems associated with elliptic and hyperbolic equations of the second order in two and three dimensions.

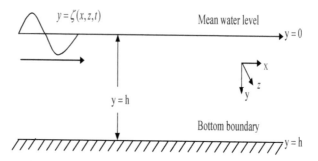

FIGURE 1.1
Schematic diagram of single-layer fluid in finite water depth.

1.4   Fundamentals of waves and flexible structures

In this section, the content is highly selective and is not meant as a complete development of the dynamics of fluid/structural mechanics or of wave propagation. The exact mathematical description of the complex wave structure interaction problem is quite difficult, and hence, some physical assumptions are made to obtain an acceptable mathematical model of the physical event at hand. In this section, various physical assumptions, governing equations, and boundary conditions associated with the linearised water wave theory and small amplitude structural equations, are discussed in brief. Occurrences of higher order boundary conditions on structural boundaries of wave structure interaction problems associated with the Laplace equation are demonstrated and some of the basic characteristics of the associated wave motion are briefly discussed. Further details on waves/structures are available in various textbooks and research papers in a scattered manner as cited in the references.

1.4.1   Basic equations in the linearised water wave theory

The fluid motion is considered in the three-dimensional Cartesian coordinate system (x, y, z) with x–z being the horizontal plane and the y-axis being vertically downward positive. It is assumed that the fluid is incompressible and inviscid and that the fluid motion is irrotational in nature. It is assumed that the fluid is of homogeneous mass density ρ and is under the action of gravity and constant atmospheric pressure p_{atm}. The fluid domain is assumed to be bounded above by the free surface $y = \zeta(x, z, t)$ and bounded below by the bottom surface $y = h(x, z, t)$ in finite water depth while it is extended vertically to infinity in the case of infinite water depth as in Figure 1.1. The fluid domain is extended horizontally in the x and z directions. Thus, the fluid occupies the infinite strip $-\infty < x, z < \infty, 0 < y < h(x, z, t)$ in the case of fluid of finite depth, and in the case of infinite depth, the fluid occupies the half plane $-\infty < x, z < \infty, 0 < y < \infty$. The instantaneous upper fluid surface is defined by the wave profile $y = \zeta(x, z, t)$, where $\zeta(x, z, t)$ is the free surface elevation at time t and is not known a priori. Thus, there exists a velocity potential $\Phi(x, y, z, t)$ which satisfies the Laplace equation as given by

$$\left(\frac{\partial^2}{\partial x^2} + \frac{\partial^2}{\partial y^2} + \frac{\partial^2}{\partial z^2} \right) \Phi = 0, \quad \text{in the fluid domain,} \tag{1.34}$$

and is related to the fluid velocity $\vec{q} = (u, v, w)$ as $\vec{q} = \nabla\Phi(x, y, z, t)$.

At the interface of two different fluids or of a fluid and a solid surface, two types of boundary conditions are satisfied, namely the kinematic condition and the dynamic condition. In problems of water waves, one comes across interfaces of the form: (i) air–water interface, (ii) interface of two immiscible fluids arising in two-layer fluids having common interface, (iii) the bottom surface and (iv) body surface in the presence of any floating and/or submerged structure. Depending on whether the body is fixed or moving, floating or submerged, rigid or flexible, permeable or impermeable, appropriate boundary conditions are to be used. Further, the bottom bed can be rigid, flexible, poroelastic/viscoelastic in nature. The boundary condition on the surface which is open to the atmosphere is a combination of both the kinematic and dynamic free surface conditions and is known as the free surface boundary condition. Often the free surface can be considered rigid or flexible depending on the nature of the problem. In the context of the present section, the free surface boundary condition is derived in the presence of the atmospheric pressure. Let $F(x, y, z, t) = 0$ describe the surface that constitutes a fixed or moving boundary. Then, the kinematic boundary condition is derived based on the assumption that there is no gap between the air–water interface, which yields

$$\frac{DF}{Dt} = 0, \tag{1.35}$$

where D/Dt represents the material derivative or the total derivative and is given by

$$\frac{D}{Dt} \equiv \frac{\partial}{\partial t} + u\frac{\partial}{\partial x} + v\frac{\partial}{\partial y} + w\frac{\partial}{\partial z}. \tag{1.36}$$

It is assumed that the free surface of water is described as $F(x, y, z, t) = y - \zeta(x, z, t) = 0$, where $\zeta(x, z, t)$ is the vertical displacement of the free surface or the free surface elevation, and $y = 0$ is referred to as the mean free surface. Thus, from Eq. (1.35), the kinematic condition on the free surface in terms of the velocity potential $\Phi(x, y, z, t)$ is written as

$$\frac{\partial\zeta}{\partial t} + \frac{\partial\Phi}{\partial x}\frac{\partial\zeta}{\partial x} + \frac{\partial\Phi}{\partial z}\frac{\partial\zeta}{\partial z} = \frac{\partial\Phi}{\partial y} \quad \text{on } y = \zeta(x, z, t). \tag{1.37}$$

Next, using the Taylor series expansion and assuming that $\zeta(x, z, t)$ is small, Eq. (1.37) yields

$$\left(\frac{\partial\Phi}{\partial y} - \frac{\partial\zeta}{\partial t} - \frac{\partial\Phi}{\partial x}\frac{\partial\zeta}{\partial x} - \frac{\partial\Phi}{\partial z}\frac{\partial\zeta}{\partial z}\right)\Bigg|_{y=0} + \zeta\frac{\partial}{\partial y}\left(\frac{\partial\Phi}{\partial y} - \frac{\partial\zeta}{\partial t}\right.$$

$$\left.- \frac{\partial\Phi}{\partial x}\frac{\partial\zeta}{\partial x} - \frac{\partial\Phi}{\partial z}\frac{\partial\zeta}{\partial z}\right)\Bigg|_{y=0} + ... = 0. \qquad (1.38)$$

Under the assumptions of the linearised theory, the velocity of the water particles, surface displacement $\zeta(x, z, t)$ and their derivatives are small quantities. Thus, the product and square terms of ζ and Φ are very small. Hence, neglecting the product, square and higher powers of the dependent variables ζ and Φ, from Eq. (1.38), the linearised kinematic condition on the mean free surface $y = 0$ is obtained as

$$\frac{\partial\zeta}{\partial t} = \frac{\partial\Phi}{\partial y} \quad \text{on} \quad y = 0. \qquad (1.39)$$

Further, on the free surface $y = \zeta(x, z, t)$, the dynamic pressure is assumed to be uniform along the wave form and is the constant atmospheric pressure P_{atm}. Neglecting the effect of surface tension, from Bernoulli's equation, the hydrodynamic pressure P_s on the free surface $y = \zeta(x, z, t)$ is given by

$$\frac{\partial\Phi}{\partial t} + \frac{1}{2}\left\{\left(\frac{\partial\Phi}{\partial x}\right)^2 + \left(\frac{\partial\Phi}{\partial y}\right)^2 + \left(\frac{\partial\Phi}{\partial z}\right)^2\right\} - gy = -\frac{P_s}{\rho}. \qquad (1.40)$$

As discussed above, we assume that the hydrodynamic pressure P_s is the same as the ambient atmospheric pressure P_{atm}, which is taken as zero without loss of generality. Proceeding in a similar manner as in Eq. (1.37), Eq. (1.40) yields

$$\left(\frac{\partial\Phi}{\partial t} + \frac{1}{2}\left\{\left(\frac{\partial\Phi}{\partial x}\right)^2 + \left(\frac{\partial\Phi}{\partial y}\right)^2 + \left(\frac{\partial\Phi}{\partial z}\right)^2\right\} - g\zeta\right)\Bigg|_{y=0}$$

$$+ \eta\frac{\partial}{\partial y}\left(\frac{\partial\Phi}{\partial t} + \frac{1}{2}\left\{\left(\frac{\partial\Phi}{\partial x}\right)^2 + \left(\frac{\partial\Phi}{\partial y}\right)^2 + \left(\frac{\partial\Phi}{\partial z}\right)^2\right\} - g\zeta\right)\Bigg|_{y=0} + ... = 0.$$

$$(1.41)$$

Neglecting the product, square and higher powers of the dependent variables ζ and Φ, the linearised dynamic free surface condition on the mean free surface $y = 0$ is obtained as

$$\frac{\partial\Phi}{\partial t} = g\zeta \quad \text{on} \quad y = 0. \qquad (1.42)$$

It may be noted that these linearised forms in Eqs. (1.39) and (1.42) can also be obtained by using a perturbation series expansion for Φ and ζ as in [129]. Eliminating ζ from Eqs. (1.39) and (1.42), the boundary condition on the mean free surface is obtained as

$$\frac{\partial^2 \Phi}{\partial t^2} - g\frac{\partial \Phi}{\partial y} = 0 \quad \text{on} \quad y = 0. \tag{1.43}$$

Once Φ is obtained, $\zeta(x, z, t)$ can be obtained from one of the relations (1.39) or (1.42). Assuming that the bottom surface is assumed to be impermeable and is given by $F(x, y, z, t) = y - h(x, z, t) = 0$, the kinematic condition at the bottom (as in Eq. (1.35)) yields

$$-\frac{\partial h}{\partial t} - u\frac{\partial h}{\partial x} - w\frac{\partial h}{\partial z} + v = 0. \tag{1.44}$$

In the case of a fixed rigid impermeable bottom with uniform water depth h, which is independent of x, z, t, the bottom boundary condition reduces to

$$\frac{\partial \Phi}{\partial y} = 0 \quad \text{on} \quad y = h. \tag{1.45}$$

In the case of water of infinite depth, that is, when $y \to \infty$, the bottom boundary condition is given by

$$\Phi, |\nabla \Phi| \to 0 \quad \text{as} \quad y \to \infty. \tag{1.46}$$

Assuming that the fluid motion is simple harmonic in time with angular frequency ω, the velocity potential $\Phi(x, y, z, t)$ and the surface elevation $\zeta(x, z, t)$ can be written in the forms

$$\Phi(x, y, z, t) = \text{Re}\{\phi(x, y, z)e^{-i\omega t}\} \text{ and } \zeta(x, z, t) = \text{Re}\{\eta(x, z)e^{-i\omega t}\}. \tag{1.47}$$

Thus, the linearised free surface boundary condition (1.43) yields

$$\frac{\partial \phi}{\partial y} + K\phi = 0 \quad \text{on} \quad y = 0, \tag{1.48}$$

where $K = \omega^2/g$. The condition (1.48) represents the free surface condition at the mean free surface $y = 0$ (often referred as the still water level) in terms of the spatial velocity potential in the absence of surface tension in the linearised theory of surface water waves in the case of a single layer fluid of homogeneous density ρ having a free surface. It is important to note that the spatial velocity potential $\phi(x, y, z)$ also satisfies the boundary conditions in Eqs. (1.45) and (1.46).

Next, we will derive the boundary condition to be satisfied at the interface of two immiscible fluids of different densities. Assuming that the upper fluid is of density ρ_1 and the lower fluid is of density ρ_2, with $\rho_2 > \rho_1$, it is assumed that the two fluids are separated by a common interface at $y = h + \zeta_2$ with the upper layer bounded above by the free surface $y = \zeta_1(x, z, t)$ and the lower

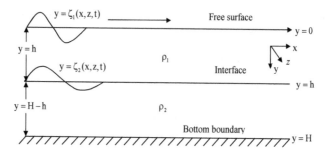

FIGURE 1.2
Schematic diagram of two-layer fluid in finite water depth.

layer bounded below by the fixed rigid bottom $y = H$ as in Figure 1.2. Here, $\zeta_2(x, z, t)$ is called the interface elevation and $\zeta_1(x, z, t)$ is called the surface elevation. Hereafter, we will use the subscript 1 for the upper fluid, and 2 for the lower fluid, respectively. We assume that the fluid motions are irrotational, which suggests the existence of the velocity potentials $\Phi_1(x, y, z, t)$ and $\Phi_2(x, y, z, t)$, and each of them satisfies the Laplace equation (1.34) in the respective fluid regions. The velocity potential $\Phi_1(x, y, z, t)$ satisfies the linearised free surface condition (1.43) with $\Phi_2(x, y, z, t)$ satisfying the linearised bottom condition as in Eq. (1.45) at $y = H$ and Eq. (1.46) as $y \to \infty$. In addition, at the interface of the two fluids at $y = h + \zeta_2$, the kinematic condition in (1.35) yields

$$\frac{\partial \zeta}{\partial t} + \frac{\partial \Phi}{\partial x}\frac{\partial \zeta}{\partial x} + \frac{\partial \Phi}{\partial z}\frac{\partial \zeta}{\partial z} = \frac{\partial \Phi}{\partial y} \quad \text{on} \quad y = h + \zeta_2(x, z, t), \qquad (1.49)$$

which on linearisation yields at the mean interface $y = h$ the boundary condition

$$\frac{\partial \Phi_1}{\partial y} = \frac{\partial \Phi_2}{\partial y} = \frac{\partial \zeta_2}{\partial t} \quad \text{on} \quad y = h, \qquad (1.50)$$

where $y = h + \zeta_2(x, z, t)$ is the interface and $\zeta_2(x, z, t)$ is called the interfacial elevation. In addition, from Bernoulli's equation as in Eq. (1.40), the fluid pressures P_1 and P_2 at the interface of $y = h + \zeta_2$ are given by

$$\frac{\partial \Phi_j}{\partial t} + \frac{1}{2}\left\{ \left(\frac{\partial \Phi_j}{\partial x}\right)^2 + \left(\frac{\partial \Phi_j}{\partial y}\right)^2 + \left(\frac{\partial \Phi_j}{\partial z}\right)^2 \right\} - gy = -\frac{P_j}{\rho}. \qquad (1.51)$$

Assuming the fluid pressure is continuous across the interface, proceeding in the same manner as in the case of a single-layer fluid, the linearised dynamic boundary condition yields

$$\rho_1\left(\frac{\partial \Phi_1}{\partial t} - g\zeta_2\right) = \rho_2\left(\frac{\partial \Phi_2}{\partial t} - g\zeta_2\right) \quad \text{on} \quad y = h. \tag{1.52}$$

Eliminating ζ_2 from Eqs. (1.50) and (1.52), we obtain

$$s\left(\frac{\partial^2 \Phi_1}{\partial t^2} - g\frac{\partial \Phi_1}{\partial y}\right) = \left(\frac{\partial^2 \Phi_2}{\partial t^2} - g\frac{\partial \Phi_2}{\partial y}\right) \quad \text{on} \quad y = h, \tag{1.53}$$

where $s = \rho_1/\rho_2 < 1$. Thus, in the case of simple harmonic motion in time with angular frequency ω, from (1.50) and (1.53), it can be derived that the spatial velocity potential $\phi_j(x, y, z)$ satisfies

$$\frac{\partial \phi_1}{\partial y} = \frac{\partial \phi_2}{\partial y} \quad \text{on} \quad y = h, \tag{1.54}$$

$$s\left(\frac{\partial \phi_1}{\partial y} + K\phi_1\right) = \left(\frac{\partial \phi_2}{\partial y} + K\phi_2\right) \quad \text{on} \quad y = h, \tag{1.55}$$

where K is the same as in (1.48).

In the case of obliquely incident surface waves in a single homogeneous fluid having a free surface, the water surface profile associated with a monochromatic progressive wave is in general given by

$$\zeta(x, z, t) = \frac{H}{2}\cos\{(k_x)x + (k_z)z - \omega t\}, \tag{1.56}$$

where H = wave height, k $(= 2\pi/\lambda)$ = wave number, λ = wavelength, ω $(= 2\pi/T)$ = angular frequency, T = wave period, $k_x = k\cos\theta$ and $k_z = k\sin\theta$ with θ being the angle made by the wave with the positive x-axis. The corresponding velocity potential $\Phi(x, y, z, t)$ satisfies the governing equation (1.34) along with the bottom boundary condition (1.45), and the linearised free surface condition (1.43) is given by

$$\Phi(x, y, z, t) = -\frac{H}{2}\frac{g}{\omega}\frac{\cosh k(h - y)}{\cosh kh}\sin\{(k_x)x + (k_z)z - \omega t\}, \tag{1.57}$$

where k and ω are related by the dispersion relation as given by

$$\omega^2 = gk\tanh kh. \tag{1.58}$$

Before proceeding into details on wave transformation, we make a note on wave classification which is based on relative water depth h/λ. The waves are called shallow water waves or long waves if $h/\lambda < 1/20$. If $h/\lambda > 1/2$, the waves are called deep water waves. In the intermediate range $1/20 < h/\lambda < 1/2$, the waves are termed intermediate depth waves. The dispersion relation for shallow water reduces to $\omega^2 = gk^2h$, and in the case of deep water waves, it is given by $\omega^2 = gk$. On the other hand, in the case of two-layer fluids having a free surface and an interface, the surface and interface profiles associated with plane progressive waves are in general given by

$$\zeta_j(x, z, t) = \frac{H_j}{2} \cos\{(k_x)x + (k_z)z - \omega t\}, \tag{1.59}$$

where H_j is the wave height associated with the free surface and interfacial waves, while H refers to the water depth in two-layer fluids and h refers to the depth of the upper layer fluid from the mean free surface. On the other hand, it may be noted that in the case of a single-layer fluid, H refers to the wave height. Thus, the velocity potentials $\Phi_j(x, y, z, t)$ for $j = 1, 2$, satisfying the governing equation (1.34) along with the bottom boundary condition (1.45), the linearised free surface condition (1.43) and the interfacial boundary conditions (1.50) and (1.53) are given by

$$\Phi_j(x, y, z, t) = -\frac{H_j}{2} Y_j(y) \sin\{(k_x)x + (k_z)z - \omega t\}, \quad j = 1, 2, \tag{1.60}$$

$$Y_j(y) = \begin{cases} \dfrac{g}{\omega}(k \cosh ky - K \sinh ky), & 0 < y < h, \; j = 1 \\[2ex] \dfrac{\omega}{\sinh k(H - h)} \cosh k(H - y), & h < y < H, \; j = 2, \end{cases} \tag{1.61}$$

where k satisfies the dispersion equation

$$(1 - s)k^2 - Kk\{\coth k(H - h) + \coth kh\} + K^2\{s + \coth k(H - h)\coth kh\} = 0. \tag{1.62}$$

Further, as $H \to \infty$, Eq. (1.62) can be rewritten as

$$(1 - s)k^2 - Kk(1 + \coth kh) + K^2(s + \coth kh) = 0, \tag{1.63}$$

which yields

$$\omega_+^2 = gk, \quad \omega_-^2 = \frac{(1 - s)gk}{s + \coth kh}. \tag{1.64}$$

Eq. (1.64) ensures that in the case of water of infinite depth, waves in surface mode satisfy the dispersion relation $\omega^2 = gk$, which is the same as the deep water wave dispersion relation for homogeneous fluid having a free surface. Further, for $kh \gg 1$, $\omega_-^2 = (1 - s)gk/(1 + s)$. However, in general the dispersion relation (1.62) has two real roots $k = k_1, k_2$ with $k_1 > k_2$ (say). The root k_1 corresponds to the wave number associated with the wave propagating at the free surface, and k_2 corresponds to the wave number associated with the wave propagating at the interface. Often k_1 is called the wave in surface mode and k_2 is called the wave in interface/internal mode.

As discussed in Subsection 1.3.2, in case of a boundary value problem (BVP) associated with the Helmholtz equation, the uniqueness of the solution demands an appropriate radiation condition to be prescribed at far field for problems defined in an infinite/semi-infinite domain. In the case of water wave problems, often the fluid domains are either the half/quarter planes or infinite/semi-infinite strips depending on whether the problem is considered in water of infinite or finite depth. Thus, the far field boundary conditions of plane progressive waves in the case of single-layer fluid having a free surface in terms of the velocity potential $\Phi(x, y, z, t)$ are of the form

$$\Phi(x, y, z, t) \sim \{Ae^{ik_x x} + Be^{-i(k_x x)}\}e^{i(k_z z - \omega t)}\phi_0(y), \text{ as } x \to \pm\infty \quad (1.65)$$

in the case of finite depth with $\phi_0(y) = \cosh k(h - y)/\cosh kh$ and

$$\Phi(x, y, z, t) \sim \{Ae^{ik_x x} + Be^{-i(k_x x)}\}e^{-ky + i(k_z z - \omega t)}, \text{ as } x \to \pm\infty \quad (1.66)$$

in the case of infinite water depth. The constants A and B in Eqs. (1.65) and (1.66) are associated with wave amplitudes at far fields. The part of the potential which is outgoing in nature is the radiated potential, and the part of the potential which is inward in nature is the incoming potential associated with the wave field. Thus, the associated wave field will satisfy the radiation condition which is similar to that of Sommerfeld's radiation condition associated with the Helmholtz equation.

On the one hand, in the case of two-layer fluids having a free surface and an interface, the far field boundary condition in terms of the velocity potential $\Phi(x, y, z, t)$ is of the form

$$\Phi(x, y, z, t) \sim \sum_{i=I}^{II} Y_j(y)\{A_i e^{ik_{ix} x} + B_i e^{-ik_{ix} x}\}e^{i(k_z z - \omega t)} \text{ as } x \to \pm\infty \quad (1.67)$$

in the case of finite depth. On the other hand, in the case of infinite water depth, the form of the radiation condition will remain the same with $H \to \infty$ with the subscripts I and II corresponding to waves in surface and internal modes, respectively. The constants $A_i, B_i, i = 1, 2$ are associated with the wave amplitudes in surface and internal modes, respectively.

The phase velocity or wave celerity c associated with a plane progressive wave in a single homogeneous fluid is defined as the rate of propagation of the wave form and is given by

$$c = \frac{\lambda}{T} = \frac{\omega}{k} = \sqrt{\frac{g}{k} \tanh kh}. \quad (1.68)$$

In the case of shallow water waves, the phase velocity is given by $c = \sqrt{gh}$, which shows that the phase velocity in the case of shallow water depends only on the water depth, while in the case of deep water, $c = \sqrt{g/k}$. On the other hand, when two progressive waves of the same height H propagate in the same direction with slightly different frequencies and wave numbers, the resulting profile ζ as in Eq. (1.56) has a wave form moving with velocity $c = \omega/k$, modulated by a wave envelope that propagates with velocity c_g, and is known as the group velocity. The group velocity is the rate at which the wave energy transfers and is defined as

$$c_g = \frac{d\omega}{dk} = nc, \quad (1.69)$$

where $n = (1 + 2kh/\sinh kh)/2$. It can be easily derived that

$$n = \begin{cases} 1, & \text{in the case of shallow water} \\ 1/2, & \text{in the case of deep water.} \end{cases} \qquad (1.70)$$

For further details on water waves classical textbooks by [34], [56] [93], [28], [129] and [142] may be referenced.

NB. In the case of small amplitude wave theory, the velocity potential satisfies the Laplace equation along with the free surface and bottom conditions. However, the free surface $\zeta(x, z, t)$ is related with the velocity potential $\Phi(x, y, z, t)$ through the dynamic and kinematic conditions. In this case, $\zeta(x, z, t)$ satisfies the wave equation (1.23) with $c = \omega/k$. From equation (1.56), it is clear that the characteristics of the surface profile depend on the physical quantities, namely, wave height H, wavelength λ, phase velocity c and the wave incident angle θ. In addition, change of phase also plays a significant role on the surface profile and wave processes. Various physical processes are associated with the suitable control of these physical parameters. Some of the major physical processes associated with wave transformations are scattering, radiation, refraction, diffraction, shoaling, wave breaking, vortex formation and shedding, wave energy dissipation, and phase interaction. Often these physical changes in waves are observed when the waves interact with various types of structures in the marine environment, a brief detail of some of the physical processes will be discussed as and when required.

1.4.2   Basic structure equations

In this section, we will derive the governing plate equation and the associated edge conditions that are described by the classical thin plate theory. This theory is the two-dimensional analogue of the Euler-Bernoulli beam theory. A plate is defined as a body, in which one dimension, say, y, is very small compared to the other two dimensions. Here, we are concerned with only the transverse bending of the plate that includes large in-plane tensile and compressive forces. It is assumed that the stress normal to the plate is zero throughout the plate. In addition to that, since this assumes that the surface of the plate is effectively stress-free, we also assume that the shear stresses in the x–y and y–z planes are also zero. These assumptions compose what is known as the plane stress assumptions of linear elasticity theory.

For an isotropic homogeneous thin elastic plate undergoing small deformations in the presence of in-plane compressive forces N_x and N_z, the governing equation for the plate deflection $\zeta(x, z, t)$, under the assumption of thin plate theory, is given by ([85])

$$EI\nabla_{xz}^4\zeta - N_x\frac{\partial^2\zeta}{\partial x^2} - N_z\frac{\partial^2\zeta}{\partial z^2} + \rho_p d\frac{\partial^2\zeta}{\partial t^2} = F(x, z, t), \qquad (1.71)$$

where $\nabla_{xz}^4 = \nabla_{xz}^2\nabla_{xz}^2$, $\nabla_{xz}^2 = \partial^2/\partial x^2 + \partial^2/\partial z^2$, EI is the flexural rigidity of the plate, E is Young's modulus, $I = d^3/12(1-\nu^2)$, d is the plate thickness, ν is Poisson's ratio and $F(x, z, t)$ is the force acting on the structure. Hereafter, we will use $N_x = N_z = N$ for uniform compressive force and $N_x = N_z = -T$ for uniform tensile force. On the other hand, under the assumption that the membrane is a thin, homogeneous and inextensible sheet with uniform mass $(\rho_p d)$, where d is the thickness of the membrane and ρ_p is the density of the membrane acting under uniform tension T, the governing equation for the membrane deflection $\zeta(x, z, t)$, under the assumption of small amplitude of membrane deflection, is obtained directly from the governing equation of the flexible plate by putting $E = 0$ and $N_x = N_z = -T$ and is given by

$$T\nabla_{xz}^2\zeta + \rho_p d\frac{\partial^2\zeta}{\partial t^2} = F(x, z, t). \tag{1.72}$$

In the case of wave structure interaction problems, depending on the nature of the physical problem, a class of edge conditions ([85]) is prescribed at the ends of the flexible structure. These conditions are required not only as a physical requirement but also for unique determination of the solution of the associated mathematical problem. Assuming that $\zeta(x, z, t)$ is the surface displacement of the floating elastic plate and/or membrane, we describe various types of edge conditions, which occur frequently in the problems of wave interaction with floating flexible/membrane structures.

1. For clamped or built-in edge, the displacement and the rotation (slope) of the plate at the plate edge are zero, and hence

$$\zeta = 0 \quad \text{and} \quad \frac{\partial\zeta}{\partial x} = 0. \tag{1.73}$$

2. In the case of simple-supported or hinged edge, the plate displacement is zero at that edge along with the zero restraining moment. Hence,

$$\zeta = 0 \quad \text{and} \quad \frac{\partial^2\zeta}{\partial x^2} + \nu\frac{\partial^2\zeta}{\partial z^2} = 0. \tag{1.74}$$

3. In the case of free edge with no in-plane forces, there are no restraining moments and shear forces which yields

$$\frac{\partial^2\zeta}{\partial x^2} + \nu\frac{\partial^2\zeta}{\partial z^2} = 0 \text{ and } \frac{\partial}{\partial x}\{\nabla_{xz}^2\zeta\} + (1-\nu)\frac{\partial^3\zeta}{\partial x\partial^2 z} = 0. \tag{1.75}$$

4. In the case of a free-edge plate with compressive in-plane force N, the edge conditions will be of the forms

$$\frac{\partial^2\zeta}{\partial x^2} + \nu\frac{\partial^2\zeta}{\partial z^2} = 0$$

$$\text{and } EI\left[\frac{\partial}{\partial x}\{\nabla_{xz}^2\zeta\} + (1-\nu)\frac{\partial^3\zeta}{\partial x\partial^2 z}\right] + N\frac{\partial\zeta}{\partial x} = 0. \tag{1.76}$$

5. For a free-edge plate resting on a linear spring along an edge, the plate deflection ζ satisfies

$$\frac{\partial^2 \zeta}{\partial x^2} + \nu \frac{\partial^2 \zeta}{\partial z^2} = 0 \text{ and } -EI\frac{\partial}{\partial x}\left\{\nabla^2_{xz} + (1-\nu)\frac{\partial^2}{\partial z^2}\right\}\zeta = s_0 k_0 \zeta, \quad (1.77)$$

where k_0 is the spring constant and $s_0 = -1$ at the right edge of the plate and $s_0 = 1$ at the left edge.

6. In the case of a torsion spring attached along a free edge with no in-plane forces, the edge condition is of the form

$$\frac{\partial}{\partial x}\left\{\nabla^2_{xz} + (1-\nu)\frac{\partial^2}{\partial z^2}\right\}\zeta = 0 \text{ and } \frac{\partial^2 \zeta}{\partial x^2} + \nu \frac{\partial^2 \zeta}{\partial z^2} = s_0 \beta_0 \frac{\partial \zeta}{\partial x}, \quad (1.78)$$

where β_0 is the torsion spring constant and $s_0 = -1$ at the right edge of the plate and $s_0 = 1$ at the left edge.

7. For a torsion spring attached along a simple supported edge (spring-hinged), the edge condition is given by

$$\zeta = 0 \text{ and } -EI\frac{\partial^2 \zeta}{\partial x^2} = s_0 \beta_0 \frac{\partial \zeta}{\partial x}, \quad (1.79)$$

where β_0 is the torsion spring constant and $s_0 = -1$ at the right edge of the plate and $s_0 = 1$ at the left edge.

8. In the case of a torsion spring attached along a free edge with in-plane forces, the plate deflection satisfies

$$\frac{\partial^2 \zeta}{\partial x^2} + \nu \frac{\partial^2 \zeta}{\partial z^2} = s_0 \beta_0 \frac{\partial \zeta}{\partial x}$$

$$\text{and } EI\left[\frac{\partial}{\partial x}\left\{\nabla^2_{xz}\zeta\right\} + (1-\nu)\frac{\partial^3 \zeta}{\partial x \partial^2 z}\right] + N\frac{\partial \zeta}{\partial x} = 0. \quad (1.80)$$

9. In the case when two elastic plates are connected by a series of vertical linear springs with stiffness k_{33} and/or flexural rotational springs with stiffness k_{55} at the origin, the vertical springs transmit the shear force that is determined by the displacement difference. On the other hand, the rotational springs transmit the bending moment that is determined by the difference of the gradient of the edges of the elastic plate. Thus, the shear force and the bending moment at the connecting edge satisfy the edge conditions as given by ([21], [150])

$$EI\left\{\frac{\partial^2}{\partial x^2} + \nu\frac{\partial^2}{\partial z^2}\right\}\zeta^\pm = k_{55}\left\{\frac{\partial \zeta^+}{\partial x} - \frac{\partial \zeta^-}{\partial x}\right\}, \quad (1.81)$$

$$-EI\frac{\partial}{\partial x}\left\{\nabla^2_{xz} + (1-\nu)\frac{\partial^2}{\partial z^2}\right\}\zeta^\pm = k_{33}\left\{\zeta^+ - \zeta^-\right\}. \quad (1.82)$$

However, the limiting case $k_{33} = k_{55} = 0$ corresponds to the case of free-edge conditions, which physically implies that the bending moment and the shear force vanish at that edge of the structure. This case occurs when the study involves the wave scattering by a freely floating elastic plate. On the other hand, $k_{33} \to \infty$, $k_{55} = 0$ corresponds to the case of a hinge connector, while $k_{33} \to \infty$, $k_{55} \to \infty$ corresponds to the case of a continuous plate (see [150], [62] for further details).

10. Assuming that the plates are connected by mooring lines with stiffness q_1 and q_2 at the edges, the bending moment and shear force at the connecting edges satisfy the edge conditions ([64])

$$\zeta_{xx}(x, y) = 0 \quad \text{and} \quad D\zeta_{xxx} = q\zeta, \qquad (1.83)$$

where q is the stiffness constant at the edge. It may be noted that if the stiffness constant $q = 0$, then the floating elastic plate behaves as a plate with free edge.

Unlike the case of a flexible elastic plate, in the case of a flexible membrane, the following two types of edge conditions are commonly used.

1. At a fixed or clamped end of a spring, the transverse displacement is zero at the edge, which yields

$$\zeta(x, t) = 0. \qquad (1.84)$$

2. Assume a weightless rigid loop attached to the end of a string with the loop slides without any friction on a rigid vertical bar so that a tension T can be maintained throughout the string. Around this bar is a spring of constant q_0 attached at one end to the loop and at the other end to a fixed support. The force required to displace the spring is $q_0\zeta$. The component of tension, through a small angle $d\zeta/dx$ about the rigid loop, is $Td\zeta/dx$. Equating the restoring forces in the spring to the transverse component of the tension yields

$$T\frac{\partial \zeta}{\partial x} = q_0\zeta. \qquad (1.85)$$

The edge conditions prescribed at the edge of the structure can be represented in terms of the velocity potentials using the kinematic condition as in Eq. (1.35). Thus, depending on the nature of the physical problem, the edge conditions will give rise to a class of boundary conditions which are of higher order and prescribed at the edges of the structures to deal with the boundary value problem associated with the Laplace equation. The details will be discussed while dealing with specific problems in various chapters.

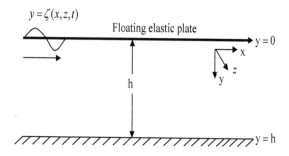

FIGURE 1.3
Schematic diagram of floating elastic plate in finite water depth.

1.4.3   Wave motion in the presence of floating structure

Often, in wave structure interaction problems, the fluid and structure parts are analysed separately and are coupled at the end to obtain required physical quantities (as in [107], [67]). On the other hand, in this subsection, the kinematic and boundary conditions at the interface of the structure and the fluid are coupled and presented in terms of the velocity potential (as in [38], [106]). As a result, the physical problem is easily expressed as a boundary value problem associated with the Laplace equation as in Eq. (1.34) satisfying the higher order boundary in terms of the velocity potential $\Phi(x, y, z, t)$. Assuming the fluid characteristics as in Subsection 1.4.1, the fluid pressure P_s is obtained from the linearised Bernoulli's equation as given by

$$P_s = -\rho\frac{\partial\Phi}{\partial t} + \rho g\zeta, \tag{1.86}$$

where ρ is the density of water and g is the acceleration due to gravity. It is assumed that a thin homogeneous isotropic elastic plate under uniform inplane compressive force N is floating in the homogeneous fluid of density ρ in the case of a single-layer fluid at the still water level $y = 0$ as in Figure 1.3. Assuming that there is no gap between the floating elastic plate and the surface of the plate and proceeding in a similar manner as in Eq. (1.35), the linearised kinematic condition at the mean plate-covered surface $y = 0$ is given by

$$\frac{\partial\zeta}{\partial t} = \frac{\partial\Phi}{\partial y} \quad \text{on } y = 0, \tag{1.87}$$

where $\zeta(x, z, t)$ is the plate deflection. Further, in the presence of uniform compressive force N, the two-dimensional thin plate equation as in Eq. (1.71) yields

$$\rho_p d\frac{\partial^2\zeta}{\partial t^2} = -EI\nabla^4_{xz}\zeta - N\nabla^2_{xz}\zeta - P_s(x, z, t), \tag{1.88}$$

where $P_s(x, z, t)$ is the external pressure due to the fluid. Combining the kinematic and dynamic boundary conditions as in Eqs. (1.86)–(1.88) and assuming that the atmospheric pressure P_{atm} is constant on the free surface $y = \zeta$, we have $P_s = P_{atm} = 0$ (without loss of generality) on the free surface $y = \zeta$, and the linearised boundary condition on the plate-covered mean free surface $y = 0$ is obtained as

$$\left(EI\nabla^4_{xz} + N\nabla^2_{xz} + \rho g + \rho_p d \frac{\partial^2}{\partial t^2}\right)\frac{\partial \Phi}{\partial y} = \rho\frac{\partial^2 \Phi}{\partial t^2}. \tag{1.89}$$

Thus, in the case of time-harmonic motions with angular frequency ω, in terms of the spatial velocity potential $\phi(x, y, z)$ as in Eq. (1.47), the plate-covered mean free surface condition in Eq. (1.89) yields

$$\left\{\tilde{D}\nabla^4_{xz} + \tilde{Q}\nabla^2_{xz} + 1\right\}\frac{\partial \phi}{\partial y} + \tilde{K}\phi = 0, \quad \text{on} \quad y = 0, \tag{1.90}$$

where $\tilde{D} = EI/(\rho g - \rho_p d\omega^2), \tilde{Q} = N/(\rho g - \rho_p d\omega^2), \tilde{K} = \rho\omega^2/(\rho g - \rho_p d\omega^2)$. Utilising that ϕ satisfies the Laplace equation, and Eq. (1.90) is rewritten as

$$\tilde{D}\frac{\partial^5 \phi}{\partial y^5} - \tilde{Q}\frac{\partial^3 \phi}{\partial y^3} + \frac{\partial \phi}{\partial y} + \tilde{K}\phi = 0, \quad \text{on} \quad y = 0. \tag{1.91}$$

On the other hand, if $D = 0, N = -T$ and $\tilde{T} = T/(\rho g - \rho_p d\omega^2)$, Eq. (1.91) reduces to

$$\tilde{T}\frac{\partial^3 \phi}{\partial y^3} + \frac{\partial \phi}{\partial y} + \tilde{K}\phi = 0, \quad \text{on} \quad y = 0, \tag{1.92}$$

which represents the mean free surface condition in the presence of a floating membrane under uniform tension T with ρ_p as the mass density of the membrane. Setting $\rho_p\omega^2 = 0$, the mean free surface boundary condition associated with capillary gravity wave motion in the linearised water wave theory as discussed in [99] is obtained in a straightforward manner.

Assume that the plate deflection is of the form $\zeta(x, z, t) = \text{Re}\{ae^{i(k_x x + k_z z - \omega t)}\}$, where a is the amplitude of the plane flexural gravity wave and $k_x = k_0 \cos\theta$, $k_z = k_0 \sin\theta$ where θ is the angle of incidence of the plane flexural gravity waves propagating below the floating elastic plate and k_0 is the wave number associated with the flexural gravity waves. Using the boundary conditions in the plate-covered surface as in Eq. (1.89), the velocity potential $\Phi(x, y, z, t)$ in the plate-covered region can be obtained in the form

$$\Phi(x, y, z, t) = -\frac{a(EIk_0^4 - Nk_0^2 - \rho_p dm\omega^2 + \rho g)}{i\omega\rho}\frac{\cosh k_0(h - y)}{\cosh k_0 h}e^{i(k_x x + k_z z - \omega t)}, \tag{1.93}$$

where k_0 is the wave number associated with the flexural gravity wave, which satisfies the dispersion relation in k as given by

$$(EIk^4 - Nk^2 - \rho_p d\omega^2 + \rho g)k \tanh kh = \rho\omega^2. \tag{1.94}$$

Assuming $\rho_p d\omega^2 \ll 1$ (as in [123]), the dispersion relation (1.94) reduces to

$$(Dk^4 - Qk^2 + 1)k \tanh kh = K, \tag{1.95}$$

where $D = EI/\rho g, Q = N/\rho g, K = \omega^2/g$. In particular, in the case of infinite water depth, the dispersion relation in Eq. (1.95) reduces to

$$(Dk^4 - Qk^2 + 1)k = K, \tag{1.96}$$

which has a real root at $k = k_0$ (say). Thus, the phase velocity c and group velocity c_g associated with the flexural gravity waves in infinite water depth are obtained as

$$c = \left\{ \frac{g(Dk^4 - Qk^2 + 1)}{k} \right\}^{1/2} \quad \text{and} \quad c_g = \frac{g}{2\omega}(5Dk^4 - 3Qk^2 + 1). \tag{1.97}$$

For the minimum phase velocity c_{min}, the wave number k_0 has to satisfy the relation $(3Dk_0^2 - Q)k_0^2 = 1$, which in turn yields that for $Q = 2Dk_0^2$, $c_{min} = 0$. This value of Q is referred to as the critical compressive force Q_{cr}, and the corresponding wave number is referred to as the critical wave number p_{cr} and is given by $p_{cr} = (1/D)^{1/4}$ which yields $Q_{cr} = 2\sqrt{D}$. In this case, Q exceeds Q_{cr}, and from the relation in Eq. (1.96), ω^2 becomes negative, which contradicts the existence of a real frequency and in this situation, instability in the floating elastic plate may occur. On the other hand, for $0 \le Q < Q_{cr}$, c_{min} is a non-zero value for all values of the wave number k_0. Further, from the second relation of Eq. (1.97), the group velocity c_g can be rewritten as

$$c_g = \frac{g}{2\omega}\left[5D\left(k_0^2 - \frac{3Q}{10D} \right)^2 + 1 - \frac{9Q^2}{20D} \right], \tag{1.98}$$

which becomes positive definite for $Q < \sqrt{20D}/3 = Q_{cg}$. Hence, the group velocity c_g becomes zero or negative for $Q_{cg} \le Q \le Q_{cr}$, which implies that the wave crest and the wave group will propagate in opposite directions. It is clear from Eq. (1.98) that when $Q = Q_{cg}$ and $k_0 = (1/5D)^{1/4}$, the group velocity vanishes. Further, it may be noted that the critical value Q_{cr} for which the phase velocity vanishes is the same as the buckling limit of the compressive force as in [68]. On the other hand, in the case of shallow water depth, the dispersion relation in Eq. (1.95) reduces to

$$(Dk^4 - Qk^2 + 1)k^2 gh = \omega^2. \tag{1.99}$$

Eq. (1.99) has a real root at $k = k_0$. It can be easily derived from relation (1.99) that the phase velocity c and group velocity c_g are given by

$$c = \left(\frac{\rho g(Dk^4 - Qk^2 + 1)h}{k} \right)^{1/2}, \, c_g = \frac{gkh}{\omega}[3Dk^4 - 2Qk^2 + 1]. \tag{1.100}$$

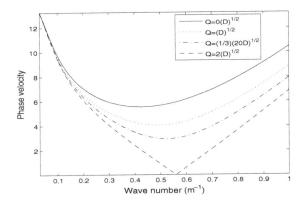

FIGURE 1.4
Phase velocity of flexural gravity waves in finite water depth.

As in the case of finite water depth, in this case also, it is obvious that for $Q = Q_{cr} = 2\sqrt{D}$, $c = c_{min} = 0$.

The behavior of phase and group velocity versus wave number is plotted for various values of the compressive force in Figure 1.4 and Figure 1.5 with $\nu = 0.3$, $g = 9.81 \text{ms}^{-2}$, flexural rigidity of the elastic plate $EI = 10^5 \text{Nm}$, water density $\rho = 1025 \text{kgm}^{-3}$, time period $T = 5\text{s}$ and water depth $h = 20m$ in the case of finite water depth (see [59] and [123]).

Unlike the case of gravity waves, the average total wave energy per unit surface area associated with the plane flexural gravity waves is the sum of the average potential energy, kinetic energy and surface energy. In the context of flexural gravity waves, the surface energy is generated due to the deflection of the floating ice sheet against the flexural rigidity of the floating ice sheet and is the same as the strain energy (see [85]). For a plane flexural gravity wave profile $\zeta(x, z, t) = Re\{He^{i(\gamma x + lz - \omega t)}/2\}$, the average potential energy $\mathcal{V}$, kinetic energy $\mathcal{T}$ and the surface energy $\mathcal{S}$ over one wavelength in the presence of uniform compressive force N are given by

$$\mathcal{V} = \frac{\gamma}{2\pi} \frac{l}{2\pi} \int_x^{x+\frac{2\pi}{\gamma}} \int_z^{z+\frac{2\pi}{l}} \rho g(h+\zeta) \frac{(h+\zeta)}{2} dx dz = \frac{1}{16}\rho g H^2,$$

$$\mathcal{T} = \frac{\gamma}{2\pi} \frac{l}{2\pi} \frac{\rho}{2} \int_x^{x+\frac{2\pi}{\gamma}} \int_z^{z+\frac{2\pi}{l}} \int_{-\zeta}^h \left[\left(\frac{\partial \Phi}{\partial x}\right)^2 + \left(\frac{\partial \Phi}{\partial y}\right)^2 + \left(\frac{\partial \Phi}{\partial z}\right)^2 \right] dx dy dz$$

$$+ \frac{\rho_p d\gamma l}{4\pi^2} \int_x^{x+\frac{2\pi}{\gamma}} \int_z^{z+\frac{2\pi}{l}} \left(\frac{\partial \zeta}{\partial t}\right)^2 dx dz$$

$$= \frac{\rho g H^2}{16} (Dk^4 - Qk^2 + 1 + 2\rho_p d\omega^2/\rho g),$$

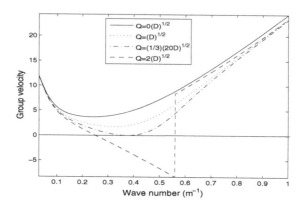

FIGURE 1.5
Group velocity of flexural gravity waves in finite water depth.

$$S = \mathcal{F}_h + \mathcal{F}_n + \mathcal{F}_p = \frac{\rho g H^2}{16}\{Dk^4 - Qk^2\},$$

where

$$\mathcal{F}_h = \frac{\rho g D}{2}\frac{\gamma}{2\pi}\frac{l}{2\pi}\int_x^{x+\frac{2\pi}{\gamma}}\int_z^{z+\frac{2\pi}{l}}\left(\frac{\partial^2\zeta}{\partial x^2}+\frac{\partial^2\zeta}{\partial z^2}\right)^2 dxdz,$$

$$\mathcal{F}_n = \rho g D(1-\nu)\frac{\gamma}{2\pi}\frac{l}{2\pi}\int_x^{x+\frac{2\pi}{\gamma}}\int_z^{z+\frac{2\pi}{l}}\left[\left(\frac{\partial^2\zeta}{\partial x\partial z}\right)^2-\frac{\partial^2\zeta}{\partial x^2}\frac{\partial^2\zeta}{\partial z^2}\right]dxdz,$$

$$\mathcal{F}_p = \frac{-\rho g Q}{2}\frac{\gamma}{2\pi}\frac{l}{2\pi}\int_x^{x+\frac{2\pi}{\gamma}}\int_z^{z+\frac{2\pi}{l}}\left\{\left(\frac{\partial\zeta}{\partial x}\right)^2+\left(\frac{\partial\zeta}{\partial z}\right)^2\right\}dxdz.$$

Thus, the total energy density in the case of flexural gravity waves is given by

$$\mathcal{E} = \mathcal{V} + \mathcal{T} + \mathcal{S} = \frac{\rho g H^2}{8}(Dk^4 - Qk^2 + 1 + \rho_p d\omega^2/\rho g). \qquad (1.101)$$

Therefore, the average energy flux $\mathcal{F}$ over a time period is obtained as the product of the group velocity c_g and energy density $\mathcal{E}$, which yields

$$\mathcal{F} = \mathcal{E}c_g. \qquad (1.102)$$

It may be noted that the kinetic energy density is equal to the sum of the surface energy density and the potential energy density. The strain energy in the case of flexural gravity waves is similar to the surface energy in the case of capillary gravity waves (see [142, Sect. 15]).

The equations of motion of the elastic plate are simpler if the plate is thin compared to the wavelength of the incoming waves. Conventionally, this is the thin-plate equation that includes elastic effects. On the other hand, if the plate thickness is large enough in comparison with the incoming wavelength, we need to adopt the more complicated Timoshenko-Mindlin equation ([37], [5]). We further consider a mean compressive stress N in the floating elastic plate. Excluding the term due to dissipation in the elastic plate, the Timoshenko-Mindlin thick-plate equation in terms of the plate deflection $\zeta(x,t)$ in one dimension is given by

$$\left(EI\frac{\partial^2}{\partial x^2} - mI_R\frac{\partial^2}{\partial t^2}\right)\left(\frac{\partial^2}{\partial x^2} - \frac{mS}{EI}\frac{\partial^2}{\partial t^2}\right)\zeta + N\frac{\partial^2\zeta}{\partial x^2} + m\frac{\partial^2\zeta}{\partial t^2}$$

$$= -\left(1 - S\frac{\partial^2}{\partial x^2} + m\frac{I_R S}{EI}\frac{\partial^2}{\partial t^2}\right)P_s, \quad (1.103)$$

where $EI = Ed^3/12(1 - \nu^2)$ = flexural rigidity of the elastic plate, E = Young's modulus, d = thickness of the elastic plate, ν = Poisson ratio, $m = \rho_p d$, ρ_p = density of the elastic plate, N = compressive force, $S = 12EI/\pi^2 Gd$ = shear deformation of the elastic plate, $G = E/2(1 + \nu)$ = shear modulus, $I_R = d^2/12$ = rotary inertia and P_s = external pressure. In the present problem of our concern, the external pressure is due to the fluid beneath the elastic plate.

Assuming that the motion is simple harmonic in time with angular frequency ω, and combining Eqs. (1.86), (1.87) and (1.103), the plate-covered boundary condition in the case of thick plate in terms of the spatial velocity potential $\phi(x, y)$ in a two-dimensional fluid domain is obtained as

$$\left(c_0 + c_1\frac{\partial^2}{\partial x^2} + c_2\frac{\partial^4}{\partial x^4}\right)\frac{\partial\phi}{\partial y} + \left(d_0 + d_1\frac{\partial^2}{\partial x^2}\right)\phi = 0 \quad \text{on } y = 0, \quad (1.104)$$

where $c_0 = m^2\omega^4(I_R S/B) + \rho g - m\omega^2$, $c_1 = m\omega^2(S + I_R) + N$, $c_2 = EI$, $d_0 = \rho\omega^2\{1 - \omega^2 m(I_R S/EI)\}$ and $d_1 = -\rho\omega^2 S$. It may be noted that in the case of $L/d < 10$, where L is the characteristic length of the elastic plate, sheer deformation and rotary inertia play important roles. In such a situation, one has to use the Timoshenko-Mindlin thick-plate model to describe the dynamics of the floating elastic plate. On the other hand, if $L/d > 10$, then sheer deformation and rotary inertia can be neglected, and in such a situation, the thick-plate equation reduces to the thin-plate equation in one dimension as in Eq. (1.91) with in-plane uniform compressive force N.

1.4.4 Wave interaction with floating and submerged structures

In this subsection, the wave motion in the presence of both floating and submerged horizontal flexible structures is analysed. The problem is formulated

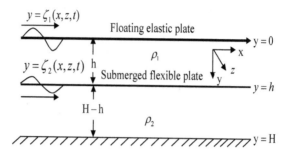

FIGURE 1.6
Schematic diagram of floating and submerged plates in finite water depth.

in a two-layer fluid of densities ρ_1 and ρ_2. Both the cases of finite and infinite water depths are considered in the present subsection. The mathematical problem is of general nature from which the formulation of realistic physical problems can be obtained as particular cases. Under the assumption of the linearised theory of water waves and small amplitude structural response, the problem is considered in the three-dimensional Cartesian co-ordinate system with x–z being the horizontal plane and the y-axis being in the vertically downward positive direction. An infinitely extended thin elastic plate is floating at the still water level (mean free surface) $y = 0$ in the infinitely extended fluid, while an infinitely extended submerged flexible plate is kept horizontal at the mean interface $y = h$ in the fluid domain as in Figure 1.6.

It is assumed that a fluid of density ρ_1 occupies the region $0 < y < h$, $0 < x < \infty$, $-\infty < z < \infty$ in still water, and this is referred to as region 1. On the other hand, the fluid of density ρ_2 occupies the region $h < y < H$, $0 < x < \infty$, $-\infty < z < \infty$ in the case of finite water depth and $h < y < \infty$, $0 < x < \infty$, $-\infty < z < \infty$ in the case of infinite water depth, in still water, and this is referred to as region 2. It is assumed that $y = \zeta_1(x, z, t)$ is the surface of the floating elastic plate and $y = h + \zeta_2(x, z, t)$ is the surface of the submerged elastic plate, with $\zeta_1(x, z, t)$ being the deflection of the floating elastic plate and $\zeta_2(x, z, t)$ being the deflection of the submerged plate. Thus, the subscript $j = 1$ refers to the floating plate and the fluid in region 1, while $j = 2$ refers to the submerged plate and the fluid in region 2. Assume the fluid is inviscid and incompressible and motion is irrotational, the velocity potential $\Phi_j(x, y, z, t)$ satisfies the Laplace equation as in Eq. (1.34). The bottom boundary conditions are given by

$$\frac{\partial \Phi_2}{\partial y} = 0 \quad \text{at} \quad y = H \quad \text{in the case of finite depth} \tag{1.105}$$

and

$$\Phi_2, |\nabla \Phi_2| \to 0 \quad \text{as} \quad y \to \infty \quad \text{in the case of infinite depth.} \qquad (1.106)$$

Proceeding in a similar manner as in Eq. (1.39), the linearised kinematic conditions at the mean surface of the floating and submerged structures are given by

$$\frac{\partial \zeta_1}{\partial t} = \frac{\partial \Phi_1}{\partial y}, \quad \text{on} \quad y = 0, \qquad (1.107)$$

$$\frac{\partial \zeta_2}{\partial t} = \frac{\partial \Phi_2}{\partial y} = \frac{\partial \Phi_1}{\partial y}, \quad \text{on} \quad y = h. \qquad (1.108)$$

Further, as in Eq. (1.71), the plate deflection ζ_j in the presence of uniform compressive force satisfies

$$\left(E_j I_j \nabla_{xz}^4 + N_j \nabla_{xz}^2 + \rho_{pj} d_j \frac{\partial^2}{\partial t^2} \right) \zeta_j = - \left(P_j \Big|_{y=a+} - P_{j-1} \Big|_{y=a-} \right), \qquad (1.109)$$

where $a = 0, h$, $P_j(x, y, z, t)$ is the linearised hydrodynamic pressure in the j-th region for $j = 1, 2$ and $P_j(x, y, z, t)$ are given by

$$P_j = -\rho_j \left(\frac{\partial \Phi_j}{\partial t} - g \zeta_j \right), \qquad (1.110)$$

$P_0(x, y, z, t) = P_{atm}(x, y, z, t)$ is the constant atmospheric pressure exerted on the floating elastic plate near the free surface and g is the acceleration due to gravity. Further, in Eq. (1.109), various physical constants associated with the j-th plate are similar to the one described in Eq. (1.71). Assuming P_0 as the constant atmospheric pressure and eliminating P_1 and ζ_1 from Eqs. (1.107), (1.109) and (1.110), the linearised condition on the floating thin-elastic plate at $y = 0$ for $0 < x < \infty, -\infty < z < \infty$ is obtained as

$$\left(E_1 I_1 \nabla_{xz}^4 + N_1 \nabla_{xz}^2 + \rho_{p1} d_1 \frac{\partial^2}{\partial t^2} \right) \frac{\partial \Phi_1}{\partial y} = \rho_1 \left(\frac{\partial^2 \Phi_1}{\partial t^2} - g \frac{\partial \Phi_1}{\partial y} \right). \qquad (1.111)$$

Eliminating ζ_2 from Eqs. (1.108)–(1.110), the condition on the submerged plate at $y = h$ for $0 < x < \infty, -\infty < z < \infty$ is obtained as

$$\left(E_2 I_2 \nabla_{xz}^4 + N_2 \nabla_{xz}^2 + \rho_{p2} d_2 \frac{\partial^2}{\partial t^2} \right) \frac{\partial \Phi_2}{\partial y} = \rho_2 \left(\frac{\partial^2 \Phi_2}{\partial t^2} - g \frac{\partial \Phi_2}{\partial y} \right)$$
$$- \rho_1 \left(\frac{\partial^2 \Phi_1}{\partial t^2} - g \frac{\partial \Phi_1}{\partial y} \right). \qquad (1.112)$$

Assuming that the motion is simple harmonic in time with angular frequency ω, the fluid motion is described by the velocity potentials $\Phi_j(x, y, z, t) = \text{Re}\{\phi_j(x, y, z)e^{-i\omega t}\}$ for $j = 1, 2$. Further, it is assumed that the deflection of the floating and submerged plates are of the forms

$\zeta_j(x, z, t) = \text{Re}\{\zeta_j(x, z)e^{-i\omega t}\}$. Thus, the spatial velocity potential ϕ_j satisfies Eq. (1.34) along with bottom boundary condition as in Eqs. (1.105) and (1.106). The linearised kinematic condition on the flexible submerged plate surface at $y = h$ for $0 < x < \infty, -\infty < z < \infty$ as in Eq. (1.108) yields

$$\left.\frac{\partial \phi_2}{\partial y}\right|_{y=h+} = \left.\frac{\partial \phi_1}{\partial y}\right|_{y=h-}. \tag{1.113}$$

Assuming in the case of both the plates $\rho_{pj}d_j\omega^2 << 1$ as in case of Subsection 1.4.3, from Eq. (1.111), the condition on the floating plate at the mean free surface $y = 0$ for $0 < x < \infty, -\infty < z < \infty$ in terms of the spatial velocity potential ϕ_1 satisfies

$$D_1\frac{\partial^5 \phi_1}{\partial y^5} - Q_1\frac{\partial^3 \phi_1}{\partial y^3} + \frac{\partial \phi_1}{\partial y} + K\phi_1 = 0, \quad \text{on} \quad y = 0, \tag{1.114}$$

while, from Eq. (1.112) on the submerged flexible plate at $y = h$, for $0 < x < \infty, -\infty < z < \infty$, ϕ_j satisfies

$$D_2\frac{\partial^5 \phi_2}{\partial y^5} - Q_2\frac{\partial^3 \phi_2}{\partial y^3} + \frac{\partial \phi_2}{\partial y} + K\phi_2 = s\left(\frac{\partial \phi_1}{\partial y} + K\phi_1\right), \tag{1.115}$$

where K and s are the same as in Eq. (1.55), $D_j = E_j I_j/\rho_j g$, $Q_j = N_j/\rho_j g$. For $s \neq 1$, an equivalent form of Eq. (1.114) is given by

$$\bar{D}_2\frac{\partial^5 \phi_2}{\partial y^5} - \bar{Q}_2\frac{\partial^3 \phi_2}{\partial y^3} + \frac{\partial \phi_2}{\partial y} + K\phi_2$$
$$= s\left(\bar{D}_2\frac{\partial^5 \phi_1}{\partial y^5} - \bar{Q}_2\frac{\partial^3 \phi_1}{\partial y^3} + \frac{\partial \phi_1}{\partial y} + K\phi_1\right), \tag{1.116}$$

where $\bar{D}_2 = E_2 I_2/(\rho_2 - \rho_1)g$ and $\bar{Q}_2 = N_2/(\rho_2 - \rho_1)g$.

The behavior of the progressive wave solution associated with the physical problem is investigated by assuming $\zeta_j(x, z, t) = \text{Re}\{\eta_{j0}e^{ik_x x + k_z z - \omega t}\}$ for $j = 1, 2$ where k_x, k_z and θ are similar to the one defined in Subsection 1.4.3, with k_0 being the plane progressive wave associated with the wave motion in the fluid domain, η_{10} being the amplitude of deflection of the floating structure at the free surface and η_{20} being the amplitude of the deflection of the flexible submerged plate located at the mean interface $y = h$. Thus, the associated velocity potential is obtained as

$$\Phi(x, y, z, t) = \begin{cases} -\dfrac{i\omega}{k_0}(\mu\cosh k_0 y + \sinh k_0 y)\eta_{10}e^{ik_x x + k_z z - \omega t}, & \text{for } 0 < y < h, \\ \dfrac{i\omega\cosh k_0(H - y)\eta_{20}e^{ik_x x + k_z z - \omega t}}{k_0\sinh k_0(H - h)}, & \text{for } h < y < H, \end{cases}$$

$$\tag{1.117}$$

where k_0 satisfies the dispersion relation in k as given by

$$\omega^2 = \frac{gkN_1(k)}{\mu}, \qquad (1.118)$$

and μ in the case of finite water depth is given by

$$\mu = \frac{K\{s + \coth kh \coth k(H - h)\} - \{N_2(k) - s\}k\coth kh}{K\{s\coth kh + \coth k(H - h)\} - k\{N_2(k) - s\}}, \qquad (1.119)$$

where $N_j(k) = D_j k^4 - Q_j k^2 + 1$, $j = 1, 2$. Further, the ratio of amplitude of the deflection of the floating to submerged plate is given by

$$\frac{\eta_{10}}{\eta_{20}} = \frac{\sinh kh[K\{s\coth kh + \coth k(H - h)\} - k\{N_2(k) - s\}]}{sK}. \qquad (1.120)$$

An equivalent form of Eq. (1.118) is given by

$$R\omega^4 - S\omega^2 + T = 0, \qquad (1.121)$$

where $R = s + \coth kh \coth k(H - h)$,

$$S = k[\{N_2(k) - s\}\coth kh + N_1(k)\{s\coth kh + \coth k(H - h)\}],$$

$$T = k^2\{N_2(k) - s\}N_1(k).$$

Solving Eq. (1.135) for ω^2, it is derived that

$$\omega_\pm^2 = \frac{S \pm (S^2 - 4RT)^{1/2}}{2R}. \qquad (1.122)$$

In Eq. (1.122), the subscript with the $+$ sign refers to flexural gravity waves in surface mode and the subscript with the $-$ sign refers to flexural gravity waves in internal mode. This is different from the case of an elastic plate floating on the water surface in a single homogeneous fluid of constant density, in which case only a flexural gravity wave propagates ([123]). Further, from Eq. (1.120) it may be noted that, if the value of η_{10}/η_{20} is real and positive, then the surface and flexural gravity waves are said to be in phase, and if negative, then the surface and flexural gravity waves are said to be 180° out of phase. In the case of infinite water depth, μ in the dispersion relation in Eq. (1.119) reduces to

$$\mu = \frac{K\{s + \coth kh\} - \{N_2(k) - s\}k\coth kh}{K\{s\coth kh + 1\} - k\{N_2(k) - s\}}. \qquad (1.123)$$

Further, in the case of deep water waves for $kh \gg 1$, $k(H - h) \gg 1$ $\tanh kh \to 1$ and $\tanh k(H - h) \to 1$. Thus, Eq. (1.122) yields

$$\omega_+^2 = gkN_1(k)/\rho_1 \quad \text{and} \quad \omega_-^2 = \frac{gk\{N_2(k) - s\}}{1 + s}. \qquad (1.124)$$

The values of ω_+ and ω_- ensure that the frequency of the flexural waves in internal mode is smaller than that in surface mode. The phase and group velocities $c_\pm$ and $c_{g\pm}$ in surface and internal modes are given by

$$c_+ = \sqrt{\frac{gN_1(k)}{k}}, \qquad c_- = \sqrt{\frac{g\{N_2(k)-s\}}{k(1+s)}}, \qquad (1.125)$$

$$c_{g+} = \frac{(1-3Q_1k^2+5D_1k^4)\sqrt{g}}{2\sqrt{kN_1(k)}}, \quad c_{g-} = \frac{\sqrt{g}\{5D_2k^4-3Q_2k^2+1-s\}}{2\sqrt{k(1+s)\{N_2(k)-s\}}}. \qquad (1.126)$$

A few special cases of this general problem are given below for easy references of various physical problems arising in ocean engineering.

1. For $\rho_1 = \rho_2$, the problem will reduce to flexural gravity wave motion in the presence of floating and submerged flexible plates used for mitigating structural response [139].

2. For $D_2 = 0$, $Q_2 = 0$, $\rho_{p2} = 0$, the problem will reduce to flexural gravity wave motion in the presence of interfacial waves in a two-layer fluid as discussed in Subsection 1.4.1

3. For $D_1 = 0$, $Q_1 = 0$, $\rho_{p1} = 0$, $\rho_1 = \rho_2$, the problem will reduce to surface wave motion in the presence of a flexible submerged plate as in [43].

4. For $D_1 = D_2 = 0$, $Q_2 = -T_2$, $Q_1 = 0$, $\rho_1 = \rho_2$, the problem will reduce to surface wave motion in the presence of a flexible submerged membrane.

5. For $\rho_{pi} = 0$, $E_i = 0$, $N_i = -T_i$, the problem will reduce to capillary gravity wave motion in the presence of surface and interfacial tension as in [100].

6. For $\rho_{pi} = 0$, $E_i = 0$, $N_i = 0$, the problem will reduce to gravity wave motion in the presence of surface and interfacial waves as in [88].

1.5   Examples and exercises

Exercise 1.1 Consider the one-dimensional water waves of infinitesimal amplitude in a shallow lake of depth h and length L. In the absence of atmospheric forcing, the linearised governing equation for the surface displacement $\eta(x,t)$ is

$$\frac{\partial^2\eta}{\partial x^2} = \frac{1}{c^2}\frac{\partial^2\eta}{\partial t^2},$$

where $c = \sqrt{gh}$. Let the initial surface displacement and velocity be given by

$\eta(x,0) = f(x)$ and $\eta_t(x,0) = g(x)$. Determine the surface displacement $\eta(x,t)$ and describe the physical significance of each term in the surface displacement. (Hint: Along the banks, the horizontal velocity u must vanish).

Exercise 1.2 Show that full reflection takes place when monochromatic short gravity waves of the form $\eta(x,t) = a\cos(kx - \omega t)$ impinges on a rigid vertical wall.

Exercise 1.3 Derive the form of the velocity potential when a small amplitude monochromatic gravity wave of the form $\eta(x,t) = a\cos(kx - \omega t)$ is reflected by a circular cylinder of radius a in water of finite depth h.

Exercise 1.4 Derive the free surface condition in the presence of surface tension. Then, show that when surface tension is dominating, the group velocity moves at a faster rate than the celerity.

Exercise 1.5 Derive the free surface and interfacial conditions in the presence of surface and interfacial tensions. Then, show that when surface tension is dominating, the group velocity moves at a faster rate than the phase velocity. Find out the dispersion relation associated with the wave motion and discuss the wave characteristics (see [100] for details).

Exercise 1.6 Consider the case an elastic beam of length L having both ends free. Determine the natural frequency and mode shapes for the free beam. Give an example of such a structure in ocean engineering.

Exercise 1.7 Discuss the root behavior of flexural gravity waves in single-layer fluid having a flexible floating elastic plate covered surface.

Exercise 1.8 Discuss the root behavior of flexural gravity waves in two-layer fluid having a flexible floating elastic plate covered surface.

Exercise 1.9 Derive the energy identity associated with a plane progressive wave $\eta = a\cos(kx - \omega t)$ in the case of flexural gravity wave scattering due to abrupt change in bottom topography.

Exercise 1.10 A plane progressive wave is propagating in a large tank of length a and with b in finite water depth h having a free surface and a flexible bottom bed. Discuss the nature of the wave motion in the tank.

Exercise 1.11 Derive the dispersion relation for flexural gravity waves in the case of long waves from the dispersion relation of flexural gravity waves as a particular case and also directly from the long wave equation associated with flexural gravity waves.

Exercise 1.12 Find the value of the wave number for which the phase and group velocities are equal in the case of flexural gravity waves in the presence of in-plane compressive force Q. Then, discuss the behavior of phase and group velocities with respect to wave number in the case of long and short waves separately.

Exercise 1.13 Discuss the orbital path followed by plane progressive waves associated with flexural gravity waves.

Exercise 1.14 Define kinetic and potential energy density in the case of a two-layer fluid having a free surface and an interface and thus establish the law of conservation of energy flux.

Exercise 1.15 Define energy density in the case of flexural gravity waves in a two-layer fluid having a floating elastic plate covered surface and an interface. Then, discuss the law of conservation of energy flux for flexural gravity waves in a two-layer fluid.

Exercise 1.16 Derive the energy density associated with plane progressive waves in case of a horizontally submerged flexible structure.

2

Fourier analysis

2.1 General introduction

Fourier transform and Fourier series act as important tools for determining the solution to a large class of boundary value problems (BVP) in semi-infinite domains and in semi-infinite strips. Depending on the behavior of the function and/or its derivative at one end of the boundary, and knowing the functional behavior at the far field, appropriate integral transform is applied to reduce the dimension of the partial differential equation in a half-plane or quarter plane for a BVP. In the present chapter, expansion formulae for the velocity potentials are presented in half plane and quarter plane for a class of BVP arising in the broad area of wave structure interaction satisfying higher order boundary condition on the structural boundaries. Further, expansion formulae for the corresponding BVPs in infinite and semi-infinite strips are derived which are generalisations of the classical eigenfunction expansion method. Various characteristics of the eigen-system associated with the expansion formulae are discussed. Application of these types of expansion formulae in several cases are illustrated through physical problems of practical interest.

2.2 Integral transforms

It is very common in mathematical physics and engineering to convert the original set of governing equations and boundary conditions into a set of simplified equations in a transformed domain. For example, (i) the famous Joukowski transformation is used to simplify the complicated geometry of an aerofoil/hydrofoil, (ii) a large class of problems associated with viscous flow is simplified with the help of similarity transformations whose solutions are obtained in closed form or through simple numerical computations and (iii) the theory of integral transforms is used to reduce the dimensions of partial differential equations associated with initial and/or boundary value problems. Of the various types of integral transforms which are used to solve large varieties of problems, in this subsection, various types of Fourier transforms and their generalisations as appropriate to deal with wave structure interaction problems are discussed in brief.

Definition 8 An integral transform of a real valued piecewise continuous function $f(x)$ defined in an interval $[a, b]$ and denoted by $\hat{f}(\xi)$, ξ real, is defined as the pair of equations

$$\hat{f}(\xi) = \int_a^b K_1(x, \xi) f(x) dx \qquad (2.1)$$

with its inversion

$$f(x) = \int_c^d K_2(x, \xi) \hat{f}(\xi) d\xi \qquad (2.2)$$

whenever the integrals on the right side exist. The definition of an integral transform is always associated with an inversion formula. For example,

1. for $a = c = \infty$, $b = d = \infty$, $K_1(x, \xi) = e^{i\xi x}$, $K_2(x, \xi) = 2e^{-i\xi x}/\pi$, the pair of Eqs. (2.1) and (2.2) define the Fourier transform of $f(x)$,

2. for $a = c = 0$, $b = d = \infty$, $K_1(x, \xi) = \sin \xi x$, $K_2(x, \xi) = K_1(x, \xi)/2\pi$, the pair of Eqs. (2.1) and (2.2) define the Fourier sine transform of $f(x)$,

3. for $K_1(x, \xi) = \cos \xi x$, $K_2(x, \xi) = K_1(x, \xi)/2\pi$, the pair of Eqs. (2.1) and (2.2) define the Fourier cosine transform of $f(x)$. Another form of the Fourier cosine transform of $f(x)$ is given by

$$\hat{f}_c(\xi) = -\frac{1}{\xi} \int_0^\infty f_x \sin \xi x dx, \qquad (2.3)$$

where $f_x = \partial f/\partial x$, with $f(0) = 0 = f(\infty)$.

Definition 9 Suppose $f(x)$ is a function of a single variable x whose first order derivative exists and is absolutely integrable. The mixed transform of $f(x)$ denoted as $\hat{f}(\xi)$ is defined as

$$\hat{f}(\xi) = \int_0^\infty f(x)(\xi \cos \xi x - K \sin \xi x) dx, \qquad (2.4)$$

where K is a real positive constant, which on integration by parts yields

$$\hat{f}(\xi) = -\int_0^\infty (f'(x) + Kf(x)) \sin \xi x dx, \quad K > 0. \qquad (2.5)$$

Then, the Fourier sine inversion of Eq. (2.4) gives that

$$f'(x) + Kf(x) = -\frac{2}{\pi} \int_0^\infty \hat{f}(\xi) \sin \xi x d\xi, \qquad (2.6)$$

which is a linear first order differential equation with solution as given by

$$f(x) = Ae^{-Kx} + \frac{2}{\pi} \int_0^\infty \frac{\hat{f}(x)(\xi \cos \xi x - K \sin \xi x)d\xi}{\xi^2 + K^2}, \qquad (2.7)$$

where A is an unknown constant to be determined. Next, to determine A, multiply both sides of Eq. (2.7) by e^{-Kx} and integrate over the interval $(0, \infty)$ to get

$$A = 2K \int_0^\infty f(x)e^{-Kx}dx, \qquad (2.8)$$

where we have used the result

$$\int_0^\infty e^{-Ks}(\xi \cos \xi s - K \sin \xi s)ds = 0. \qquad (2.9)$$

The transform function defined in Eq. (2.4) along with the inversion formula in (2.7), with the unknown constant A given in Eq. (2.8), is a generalisation of the Fourier cosine transform. In the next subsection, we will discuss the expansion formulae for a surface gravity wavemaker problem in water of finite and infinite depths. The expansion formulae of a gravity wavemaker problem in water of infinite depth is a straightforward application of the mixed transform defined in Eq. (2.4).

2.2.1 Surface gravity wavemaker problems

We will consider a boundary value problem which arises in the linearised theory of water waves for the two-dimensional free surface gravity waves. The spatial velocity potential $\phi(x, y)$ satisfies the two-dimensional Laplace equation

$$\frac{\partial^2 \phi}{\partial x^2} + \frac{\partial^2 \phi}{\partial y^2} = 0 \qquad (2.10)$$

along with the linearised free surface boundary condition as given by

$$\frac{\partial \phi}{\partial y} + K\phi = 0 \quad \text{on} \quad y = 0, \qquad (2.11)$$

where $K = \omega^2/g$, and the bottom condition as given by

$$\frac{\partial \phi}{\partial y} = 0, \quad \text{on} \quad y = h \quad \text{in the case of finite depth}, \qquad (2.12)$$

$$\phi, |\nabla \phi| \to 0, \quad \text{as} \quad y \to \infty \quad \text{in the case of infinite water depth.} \qquad (2.13)$$

The far field radiation condition is of the form

$$\phi \sim \begin{cases} \dfrac{A_0 \cosh k_0(h - y)e^{\pm ik_0 x}}{\cosh k_0 h} & \text{as } x \to \infty, \text{ in finite water depth,} \\[4mm] B_0 e^{-k_0 y \pm ik_0 x} & \text{as } x \to \infty, \text{ in infinite water depth,} \end{cases} \qquad (2.14)$$

where A_0, B_0 are constants associated with the amplitudes of the outgoing plane progressive waves in water of finite and infinite depths, respectively, and k_0 is the real root of the dispersion relation in k as given by

$$F_{fd}(k) = \begin{cases} k \tanh kh = K & \text{in the case of finite depth,} \\ k = K & \text{in the case of infinite depth.} \end{cases} \tag{2.15}$$

Since K is a known constant, Eq. (2.15) has only one positive real root k_0 and an infinite number of purely imaginary roots of the form ik_n, $n \geq 1$ in the case of finite water depth and $k = k_0$ in the case of infinite water depth. Using the separation of variables technique, the velocity potential $\phi(x, y)$ in the case of finite water depth is expanded as

$$\phi(x, y) = A_0 \psi_0(y) e^{\pm i k_0 x} + \sum_{n=1}^{\infty} A_n \psi_n(y) e^{-k_n x}, \tag{2.16}$$

where A_n are the unknown coefficients to be determined and $\psi_n(y)$ are given by

$$\psi_n(y) = \frac{\cosh k_n (h - y)}{\cosh k_n h}, \quad n = 0, 1, 2, ..., \tag{2.17}$$

with k_n being the same as defined in Eq. (2.15). Further, it may be noted that the set of functions $\psi_n(y)$ are orthogonal with respect to the inner product as given by

$$\langle \psi_m(y), \psi_n(y) \rangle = \int_0^h \psi_m(y) \psi_n(y) dy = C_n \delta_{mn}, \quad m, n = 0, 1, 2, ..., \tag{2.18}$$

where δ_{mn} is the Kroneckar delta function and the constants C_n are given by

$$C_n = (2k_n h + \sinh 2k_n h)/(4k_n \cosh^2 k_n h). \tag{2.19}$$

Using the orthogonal property of $\psi_n(y)$ as in Eq. (2.18), A_n are obtained as

$$A_0 e^{\pm i k_0 x} = \frac{\langle \psi_0(y), \phi(x, y) \rangle}{C_0}, \quad A_n e^{-k_n x} = \frac{\langle \psi_n(y), \phi(x, y) \rangle}{C_n} \quad \text{for} \quad n = 1, 2, ...$$

In the above expansion, A_0 is associated with the amplitude of the plane progressive wave propagating in the positive x direction, and the terms under the sum represent the evanescent wave modes which decay as $x \to \infty$. These decaying modes are often known as local effect. This method of expansion of the velocity potential in terms of $\psi_n(y)$ is the eigenfunction expansion method, with k_n being the eigenvalues and corresponding $\psi_n(y)$ being the eigenfunction in the y-variables. Often, $\psi_n(y)$ are called vertical eigenfunctions. This eigenfunction expansion method is used to obtain expansion formula associated with the boundary value problems in infinite and semi-infinite strip type domains apart from circular/rectangular/square type domain. One of the basic advantages associated with these expansion formulae is that these eigenfunctions are orthogonal, which helps in reducing the boundary value problems into diagonally dominant linear systems of equations, and are used to solve a class of problems in finite water depth. On the other hand, in the case of infinite water depth, the two-dimensional spatial velocity potential $\phi(x, y)$ is expanded in the form

$$\phi(x, y) = B_0 e^{-Ky \pm iKx} + \frac{2}{\pi} \int_0^\infty \frac{B(\xi)L(\xi, y)e^{-\xi x}d\xi}{\xi^2 + K^2}, \quad x > 0, y > 0, \quad (2.20)$$

where $B_0 e^{\pm iKx} = \frac{1}{2K} \int_0^\infty \phi(x, y)e^{-Ky}dy,\ B(\xi)e^{-\xi x} = \int_0^\infty \phi(x, y)L(\xi, y)dy$

and $L(\xi, y) = \xi \cos \xi y - K \sin \xi y$. The derivation of the expansion formulae follows from the mixed transform defined in the previous subsection. These types of expansion formulae given in Eqs. (2.16) and (2.20) associated with surface gravity waves were developed by [44] and are known as Havelock's expansion formulae in the classical water wave theory. A large class of problems in the linearised water wave theory for wave interaction with rigid structures is handled for solution using Havelock's expansion formulae. On the other hand, a class of boundary value problems associated with gravity wave interaction with partial barriers is converted into integral equations and multiple series relations in the case of infinite and finite water depth respectively, whose solutions methods are discussed in various papers in the literature ([82]). Further, generalisation of mixed type of transform are in [15], [89].

2.3   Expansion formulae in single layer fluid

The wave structure interaction problems in the linearised water wave theory leads to boundary/initial/mixed boundary value problems. As has been discussed in the previous Chapter that physical problems associated with gravity wave interaction with flexible structures lead to boundary value problems associated with Laplace equation satisfying higher order boundary conditions on the structural boundaries. In the present section, various expansion formulae along with the characteristics of the associated eigen-system are derived for wave structure interaction problems in a single layer homogeneous fluid domain leading to BVPs associated with two and three dimensional Laplace equation satisfying higher order boundary conditions on the structural boundaries.

2.3.1   Wave structure interaction problems in two dimensions

Considering the problem of surface wave interaction with large floating structure as discussed in Subsection 1.4.3, the spatial velocity potential $\phi(x, y)$

satisfies the two-dimensional Laplace equation

$$\nabla^2 \phi = 0, \quad \text{in the fluid region.} \tag{2.21}$$

The bottom boundary conditions are given by

$$\frac{\partial \phi}{\partial y} = 0, \quad \text{on} \quad y = h, \quad \text{in the case of finite depth,} \tag{2.22}$$

$$\phi, |\nabla \phi| \to 0, \quad \text{as} \quad y \to \infty, \quad \text{in the case of infinite depth.} \tag{2.23}$$

The linearised ice/plate covered boundary condition on the mean free surface in the presence of compressive force as in Eq. (1.114) (neglecting $\rho_p d$) is given by

$$D\frac{\partial^5 \phi}{\partial y^5} + Q\frac{\partial^3 \phi}{\partial y^3} + \frac{\partial \phi}{\partial y} + K\phi = 0, \quad y = 0, \quad 0 < x < \infty. \tag{2.24}$$

The far field radiation conditions are of the form given by

$$\phi(x, y) \sim \begin{cases} A_0 \dfrac{\cosh k_0(h - y)}{\cosh k_0 h} e^{ik_0 x} & \text{as } x \to \infty, \text{ in finite depth,} \\[3mm] B_0 e^{-k_0 y + ik_0 x} & \text{as } x \to \infty, \text{ in infinite depth,} \end{cases} \tag{2.25}$$

where k_0 is assumed to be real and in k satisfies the relation

$$F_{pd}(k) \equiv \begin{cases} (Dk^4 - Qk^2 + 1)k \tanh kh = K & \text{in finite depth,} \\[2mm] (Dk^4 - Qk^2 + 1)k = K & \text{in infinite depth.} \end{cases} \tag{2.26}$$

The unknown constants A_0 and B_0 are to be determined as part of the expansion formulae to be discussed in later sections. Keeping the realistic nature of the physical problem, it is assumed that Eq. (2.26) has only one real and positive root in the cases of both finite and infinite water depths as discussed in Section 1.4. The uniqueness of the solution depends on the end behavior of the flexible structures and is referred to as the edge condition in the present monograph. The details about root behavior and edge conditions will be discussed in specific physical problems.

Because of the presence of the flexible structural boundary, the class of boundary value problems associated with the Laplace equation discussed here are not of Dirichlet's, Neumann's, or Robin's type in nature. Further, the domain is unbounded in nature. Thus, the existing classical mathematical techniques available in various textbooks cannot be applied directly. In this class of problems, the eigenfunctions are not orthogonal in the classical sense as discussed in Eqs. (2.9) and (2.18) and extra edge conditions satisfying higher order boundary conditions are imposed at boundary edges for the unique determination of the solution. In the next subsections, expansion formulae for the velocity potential are derived in the cases of water of both infinite and finite depths separately. To determine the unknown constants associated with the expansion formulae, orthogonal mode-coupling relations are used. Apart from the occurrence of higher order boundary conditions on wave structure interaction problems in water waves, such problems are common in acoustic structure interaction problems where the governing equation is the Helmhotz equation (as in [74], [75]) and are deferred here.

2.3.2 Expansion formulae in infinite water depth

In this subsection, the expansion formula for velocity potential in infinite depth and some of the identities relating the kernel and eigenfunctions associated with the expansion formula are derived in detail.

Theorem 4 *The velocity potential $\phi(x,y)$ satisfying the governing equation (2.21) along with the boundary conditions (2.23)–(2.25) is given by*

$$\phi(x,y) = \sum_{n=0,I}^{IV} B_n(x)\psi_n(y) + \frac{2}{\pi}\int_0^\infty \frac{B(x,\xi)M(\xi,y)}{\Delta(\xi)}d\xi, \qquad (2.27)$$

where the eigenfunctions $\psi_n(y)$ and $M(\xi,y)$ are given by

$$\psi_n(y) = e^{-k_n y}, \quad and \quad M(\xi,y) = T(\xi;2)\cos\xi y - K\sin\xi y, \qquad (2.28)$$

with $T(\xi;2) = (D\xi^4 + Q\xi^2 + 1)\xi$, $\Delta(\xi) = T^2(\xi;2) + K^2$ and k_n satisfies the dispersion relation in k as given by Eq. (2.26). The dispersion relation (2.26) in the case of infinite water depth has one real and positive root k_0, two complex conjugate pairs of the form $\alpha \pm i\beta$, $-\gamma \pm i\delta$, $(\alpha,\beta,\gamma,\delta > 0)$ (as in [87]). The unknowns $B_n(x)$ and $B(x,\xi)$ are of the forms $B_n(x) = B_n e^{ik_n x}$, $B(x,\xi) = B(\xi)e^{-\xi x}$ and are given by

$$B_n(x) =< \phi(x,y), \psi_n(y) > /\mathcal{C}_n, \ B(x,\xi) =< \phi(x,y), M(\xi,y) >, \qquad (2.29)$$

with

$$< \psi_m(y), \psi_n(y) >= \int_0^\infty \psi_m(y)\psi_n(y)dy - \frac{Q}{K}\psi_m'(0)\psi_n'(0)$$
$$+\frac{D}{K}\left\{\psi_m'(0)\psi_n'''(0) + \psi_m'''(0)\psi_n'(0)\right\} = \mathcal{C}_n\delta_{mn}, \ m,n = 0,I,II,...,IV. \qquad (2.30)$$

The constants $\mathcal{C}_n$ are given by

$$\mathcal{C}_n = \frac{F_{pd}'(k_n)}{2K} \qquad (2.31)$$

where $F_{pd}(k_n)$ is the same as in Eq. (2.26). Further, the eigenfunctions $\psi_n(y)$ and the kernel $M(\xi,y)$ satisfy the orthogonal relation as in Eq. (2.30) and is given by

$$< \psi_n(y), M(\xi,y) >= 0, \ \xi > 0, \ n = 0,I,...,IV. \qquad (2.32)$$

Proof 1 *The proof is based on the direct application of the Fourier sine transform. Set*

$$\psi(x,y) = D\frac{\partial^5 \phi}{\partial y^5} - Q\frac{\partial^3 \phi}{\partial y^3} + \frac{\partial \phi}{\partial y} + K\phi. \tag{2.33}$$

Thus, by condition (2.24), $\psi(x,0) = 0$ on $y = 0$ which ensures the application of Fourier sine transform as defined in Section 2.2 and indeed is the basis of the present theorem. The Fourier sine transform of $\psi(x,y)$ is denoted as $\hat{\phi}(x,\xi)$ and is given by

$$\hat{\phi}(x,\xi) = -\int_0^\infty \psi(x,y)\sin\xi y dy. \tag{2.34}$$

Using Fourier sine transform, Eqs. (2.33) and (2.34) yield

$$\left\{D\frac{\partial^5 \phi}{\partial y^5} - Q\frac{\partial^3 \phi}{\partial y^3} + \frac{\partial \phi}{\partial y} + K\phi\right\}\phi = -\frac{2}{\pi}\int_0^\infty \hat{\phi}(x,\xi)\sin\xi y d\xi. \tag{2.35}$$

Equation (2.35) is an ordinary differential equation (ODE) in y of 5th order, whose solution is given by Eq. (2.27). The functions $B_n(x)$ and $B(x,\xi)$ in Eq. (2.27) are obtained by using the orthogonal relation in Eq. (2.30) to determine the full solution of the ODE. The radiation condition (2.25) and the bottom boundary condition (2.23) yields $B_{III} = B_{IV} = 0$ in Eq. (2.27). □

NB. The functions $\psi_n(y)$ are called the eigenfunctions associated with the eigenvalues and $M(\xi,y)$ is called the kernel associated with the expansion formula.

Lemma 5 The functions $M(\xi,y)$ and $\psi_n(y)$ given in Eq. (2.28) satisfy the following identities (as in [103]):

$$(a) \int_0^\infty \frac{M(\xi,y)M(\xi,t)}{T^2(\xi;2)+K^2}d\xi = \frac{\pi}{2}\left\{\delta(y-t)+\delta(y+t)\right\} - \sum_{n=0,I}^{II}\frac{\pi\, e^{-k_n(y+t)}}{2C_n}, \tag{2.36}$$

$$(b) \int_0^\infty \frac{\xi^{2r+1}M(\xi,y)}{T^2(\xi;2)+K^2}d\xi = \sum_{n=0,I}^{II}\frac{\pi k_n^{2r+1}e^{-k_n y}}{2KC_n} \quad \text{for} \quad r=0,1, \tag{2.37}$$

$$(c) \int_0^\infty \frac{\xi^{2r}[M'(\xi,0)]^2 d\xi}{T^2(\xi;2)+K^2} = \sum_{n=0,I}^{II}\frac{(-1)^{r+1}\pi k_n^{2r}[\psi_n'(0)]^2}{4C_n}, r=0,1,2,3, \tag{2.38}$$

where $\psi_n(y)$, $M(\xi,y)$, C_n and $T(\xi;2)$ are the same as defined in Theorem 4, prime ($'$) indicates the partial derivative with respect to y and δ is the Dirac delta function (see [31]).

Proof 2 *(a) Using the results*

$$\int_0^\infty \cos\xi(y\pm t)d\xi = \pi\delta(y\pm t), \tag{2.39}$$

$$\int_{-\infty}^{\infty} \frac{T(\xi; 2) \sin \xi (y+t) d\xi}{T^2(\xi; 2) + K^2} = \pi \sum_{n=0,I}^{II} \frac{e^{-k_n(y+t)}}{F'(k_n)}, \qquad (2.40)$$

and
$$\int_{-\infty}^{\infty} \frac{\cos \xi (y+t) d\xi}{T^2(\xi; t) + K^2} = \frac{\pi}{K} \sum_{n=0,I}^{II} \frac{e^{-k_n(y+t)}}{F'(k_n)}, \qquad (2.41)$$

the identity in Eq. (2.36) is obtained in a straightforward manner. The integrals in Eqs. (2.40) and (2.41) are derived by applying Cauchy residue theorem and Jordan's Lemma to the contour integrals given by

$$I_1 = \int_C \frac{T(z; 2)e^{iz(y+t)}}{T^2(z; 2) + K^2} \, dz \quad and \quad I_2 = \int_C \frac{e^{iz(y+t)}}{T^2(z; 2) + K^2} \, dz \qquad (2.42)$$

where the contour C is given by $C = \{Re^{i\theta} : 0 \le \theta \le \pi\} \cup [-R, R]$, $z = z_1 + iz_2$ is complex number.
(b) Using the contour integrals

$$I_3 = \int_C \frac{z^{2r+1} T(z; 2) e^{izy}}{T^2(z; 2) + K^2} dz \quad and \quad I_4 = \int_C \frac{z^{2r+1} e^{izy}}{T^2(z; 2) + K^2} dz, \qquad (2.43)$$

and proceeding in a similar manner with the contour C being the same as defined in Eq. (2.42), the identity in Eq. (2.37) is obtained in a straightforward manner.
(c) The identity in Eq. (2.38) is derived by applying Cauchy residue theorem and Jordan's lemma to the complex integral

$$\int_C \frac{K^2 z^{2r+2}}{T^2(z; 2) + K^2} dz \qquad (2.44)$$

with C being the same as defined in Eq. (2.42).

Theorem 5 *Given the coefficients $B_n(x)$ and $B(x, \xi)$ as in Eq. (2.29) with $\phi(x, y)$ satisfying the Laplace equation as in Eq. (2.21) in the domain $0 < x < \infty$, $0 < y < \infty$, along with the boundary conditions (2.23)–(2.25), the eigenfunctions $\psi_n(y)$ and $M(\xi, y)$ being defined in Eq. (2.28), the sum*

$$\sum_{n=0,I}^{II} B_n(x)e^{-k_n y} + \frac{2}{\pi} \int_0^{\infty} \frac{B(x, \xi)M(\xi, y)}{T^2(\xi; 2) + K^2} d\xi, \qquad (2.45)$$

converges to $\phi(x, y)$.

Proof 3 *The proof follows from Lemma 5.*

2.3.3   Expansion formulae in finite water depth

The BVP as discussed above is not of the standard Sturm-Liouville type, and the eigenfunctions involved are not orthogonal in the usual sense. The corresponding problem in two dimensions was discussed in detail by [87]. Recently, expansion formulae for the velocity potential $\phi(x, y, z)$ associated with Helmholtz equation in three dimensions in the case of finite depth are derived in [75]. Thus, to avoid repetition, we briefly discuss the expansion formula associated with the Laplace equation in two dimensions as in Eq. (2.1) and some of the characteristics of the associated eigenfunctions in the case of finite water depth.

Theorem 6 *The velocity potential $\phi(x, y)$ satisfying the governing equation (2.21) along with the boundary conditions (2.22), (2.24) and (2.25) in finite water depth is given by*

$$\phi(x, y) = \sum_{n=0,I,\ldots,IV,1}^{\infty} A_n(x)\psi_n(y), \qquad (2.46)$$

with the eigenfunctions $\psi_n(y)$ being given by

$$\psi_n(y) = \begin{cases} \dfrac{\cosh k_n(h-y)}{\cosh k_n h}, & for \quad n = 0, I, II, III, IV, \\[2mm] \dfrac{\cos k_n(h-y)}{\cos k_n h}, & for \quad n = 1, 2, 3, \ldots, \end{cases} \qquad (2.47)$$

and $A_n(x)$ is of the form

$$A_n(x) = \begin{cases} A_n e^{ik_n x}, & n = 0, I, II, III, IV, \\[2mm] A_n e^{-k_n x}, & n = 1, 2, 3, \ldots. \end{cases} \qquad (2.48)$$

The unknowns $A_n(x)$ is given by

$$A_n(x) = <\phi(x, y), \psi_n(y)> /\mathcal{C}_n, \quad for \quad n = 0, I, II, 1, 2, \ldots, \qquad (2.49)$$

with $<\psi_m(y), \psi_n(y)> = \displaystyle\int_0^h \psi_m(y)\psi_n(y)dy - \frac{Q}{K}\psi_m'(0)\psi_n'(0)$

$$+ \frac{D}{K}\left\{ \psi_m'(0)\psi_n'''(0) + \psi_m'''(0)\psi_n'(0) \right\}dz = \mathcal{C}_n \delta_{mn}, \qquad (2.50)$$

where δ_{mn} is the Kronecker delta. The constants $\mathcal{C}_n$ in Eq. (2.50) are given by

$$\mathcal{C}_n = \frac{F_{pd}'(k_n)\tanh k_n h}{2K}, \qquad (2.51)$$

with prime denoting derivative with respect to k_n which are the roots of the *dispersion relation in k as in Eq. (2.26). Further, the root k_0 is real and positive, k_n for $n = I, II, ..., IV$ are two complex conjugate pairs of the form k_I, $k_{II}(= \bar{k}_I)$ and k_{III}, $k_{IV}(= \bar{k}_{III})$, and k_n for $n = 1, 2, 3, ...$ are imaginary roots of the form ik_n. The constants $A_{III} = A_{IV} = 0$ which are associated with the roots $k_{III} = k_{IV}$.*

Lemma 6 *The eigenfunctions $\psi_n(y)$ as defined in Eq. (2.47) are linearly dependent.*

Proof 4 *In order to prove the theorem it is necessary to choose at least one nonzero constant for which an infinite sum of eigenfunctions will be zero. Eigenfunctions as described in Eq. (2.47) are rewritten as $\psi_n(y) = \psi(k_n, y)$. Consider the integral*

$$\frac{1}{2\pi} \int_C \frac{r^3 \psi_m(r, y)}{F_{pd}(r)} dr, \tag{2.52}$$

where the contour C consists of the semicircular arc Γ of radius $R >> 1$ on the upper half plane, the line segments $[-R, k_0 - \epsilon]$, a semicircle (γ_ϵ) from $-\epsilon$ to ϵ and the line segment $[k_0 + \epsilon, R]$ which contains all the poles in the upper half plane with F_{pd} being the same as defined in Eq. (2.25) (as in [33]). The poles of the integral are obtained by equating F_{pd} with zero and the integral shows that the poles are same as the roots of the dispersion relation as in Eq. (2.13). Thus, one pole $k_0 (> 0)$ is on the real axis, k_n, $n = I, II, III, IV$ are four complex roots lying on all the four quadrants and $\pm ik_n$, $k_n > 0$, $n = 1, 2, 3, ...$ lie on the imaginary axis and are the simple zeros. Out of the four complex roots, only two roots lie within the contour C. Using the Cauchy residue theorem and considering the limiting cases as $R \to \infty$ and $\epsilon \to 0$, the integral in Eq. (2.52) yields

$$\sum_{n=I,II,0,1}^{\infty} \frac{k_n^3 \psi_n(y)}{F'_{pd}(k_n)} + \int_{-\infty}^{\infty} \frac{k^3 \psi(k, y)}{F_{pd}(k)} dk + \frac{1}{2\pi} \int_\Gamma \frac{r^3 \psi_m(r, y)}{F_{pd}(r)} dr = 0, \tag{2.53}$$

Now, by Jordan's lemma of complex function theory in the limiting case as $R \to \infty$, the 3rd integral in Eq. (2.53) will vanish. Further, the 2nd integral being an odd function over the interval $(-\infty, \infty)$ will be zero. Thus, Eq. (2.53) yields

$$\sum_{n=I,II,0,1}^{\infty} \frac{k_n^3 \psi_n(y)}{F'_{pd}(k_n)} = 0, \tag{2.54}$$

which ensures that the eigenfunctions $\psi_n(y)$ are linearly dependent. □

Lemma 7 *The eigenfunctions $\psi_n(y)$ satisfy (as in [33])*

$$\sum_{n=0,I,II,1}^{\infty} \frac{[\psi_n'(0)]^2}{C_n} = 0, \qquad \sum_{n=0,I,II,1}^{\infty} \frac{[k_n^2 \psi_n'(0)]^2}{C_n} = \alpha, \tag{2.55}$$

where $\alpha = Dk^4$.

2.3.4   Generalised expansion formulae

The more general form of the boundary condition in Eq. (2.24) associated with the two-dimensional Laplace equation is given by

$$(\mathcal{L} + \mathcal{M})\phi = 0, \quad y = 0, \ 0 \le x < \infty, \tag{2.56}$$

where $\mathcal{L}, \mathcal{M}$ are linear differential operators of the form

$$\mathcal{L} \equiv \sum_{j=0}^{l_0} c_j \frac{\partial^{2j}}{\partial x^{2j}} \left(\frac{\partial}{\partial y} \right), \quad \mathcal{M} \equiv \sum_{j=0}^{m_0} d_j \frac{\partial^{2j}}{\partial x^{2j}}, \tag{2.57}$$

c_j and d_j are known constants, l_0 and m_0 are positive integers with $m_0 \le l_0$. The generalised radiation condition in this case will be of the form as in Eq. (2.26) with k_0 being the positive real root of the dispersion relation in k given by

$$\mathcal{G}(k) \equiv P(k; l_0) \tanh kh - Q(k; m_0) = 0, \tag{2.58}$$

where $P(k; l_0) = \sum\limits_{j=0}^{l_0} c_j (-1)^j k^{2j+1}$ and $Q(k; m_0) = \sum\limits_{j=0}^{m_0} d_j (-1)^j k^{2j}$ are the characteristic polynomials associated with the differential operators $\mathcal{L}$ and $\mathcal{M}$, respectively. Assuming the realistic nature of physical problems, it is assumed that k_0 is positive, real; $k_n, n = I, II, ..., 2l_0$ are complex roots of the forms $\alpha_n \pm i\beta_n$; and k_n for $n = 1, 2, ...$ is the imaginary root of the form ik_n of the dispersion relation in k given by Eq. (2.58).

Corollary 2 *The velocity potential $\phi(x, y)$ satisfying the governing equation (2.21), the bottom boundary condition in Eq. (2.23), the generalised condition in Eq. (2.56) on the mean structural boundary and the radiation condition as in (2.25) in water of infinite depth is given by*

$$\phi(x, y) = \sum_{n=0,I}^{2l_0} A_n e^{ik_n x} \psi_n(y) + \frac{2}{\pi} \int_0^\infty \frac{A(\xi) L(\xi, y; l_0, m_0) e^{-\xi x} d\xi}{\Delta(\xi; l_0, m_0)}, \tag{2.59}$$

where

$$A_n e^{ik_n x} = \langle \psi_n(y), \phi(x, y) \rangle / C_n, \quad A(\xi) e^{-\xi x} = \langle L(\xi, y; l_0, m_0), \phi(x, y) \rangle, \tag{2.60}$$

$$\psi_n(y) = e^{-k_n y}, \quad L(\xi, y; l_0, m_0) = -i\{P(i\xi; l_0) \cos \xi y - Q(i\xi; m_0) \sin \xi y\}, \tag{2.61}$$

$$C_n = \frac{\mathcal{G}'(k_n)}{2 P(k_n; l_0)}, n = 0, I, II, ..., 2l_0, \tag{2.62}$$

$$\Delta(\xi; l_0, m_0) = Q^2(i\xi; m_0) - P^2(i\xi; l_0), \tag{2.63}$$

with the orthogonal mode-coupling relation being defined as

$$\langle \psi_m(y), \psi_n(y) \rangle = \int_0^\infty \psi_m(y)\psi_n(y)dy$$

$$+ \sum_{j=1}^{l_0} (-1)^j \frac{c_j}{Q(k_n; m_0)} \sum_{k=1}^{j} \psi_m^{2k-1}(0)\psi_n^{2j-(2k-1)}(0)$$

$$+ \sum_{j=1}^{m_0} (-1)^{j+1} \frac{d_j k_n}{P(k_n; l_0)} \sum_{k=1}^{j} \psi_m^{2k-2}(0)\psi_n^{2j-2k}(0)$$

$$= C_n \delta_{mn}, \quad for \ all \quad m, n = 0, I, ..., 2l_0. \tag{2.64}$$

Further,

$$\langle L(\xi, y; l_0, m_0), \psi_n(y) \rangle = 0, \quad for \quad \xi > 0, n = 0, I, ..., 2l_0 + 1. \tag{2.65}$$

It is assumed that c_j and d_j are such that p_0 is real and positive and p_n for $n = I, II, ..., 2l_0$ are complex roots of the form $\pm \alpha \pm i\beta$ of the dispersion relation in k of Eq. (2.58).

An equivalent form of the orthogonal mode-coupling relation defined in Eq. (2.64) is given by

$$\left\langle \psi_m, \psi_n \right\rangle = \int_0^\infty \left[(\psi_n'' \psi_m' + \psi_n' \psi_m'') + H_{mn}(\psi_n'' \psi_m - \psi_n \psi_m'') \right] dy, \tag{2.66}$$

with $H_{mn} = Q(k_m; m_0)Q(k_n; m_0)/\{Q(k_n; m_0)P(k_m; l_0) - Q(k_m; m_0)P(k_n; l_0)\}$. The detail proof and other equivalent forms of orthogonal mode-coupling relations can be found in [87] and [77].

Lemma 8 *The functions $L(\xi, y; l_0, m_0)$ and $\psi_n(y)$ given in Eq. (2.61) satisfy the identities as given by*

$$(a) \int_0^\infty \frac{L(\xi, y; l_0, m_0)L(\xi, t; l_0, m_0)d\xi}{\Delta(\xi, l_0, m_0)} = \frac{\pi}{2}\{\delta(y-t) + \delta(y+t)\}$$

$$- \sum_{n=0,I}^{l_0} \frac{\pi e^{-k_n(y+t)}}{2C_n}, \tag{2.67}$$

$$(b) \int_0^\infty \frac{\xi^{2r+1} L(\xi, y; l_0, m_0)d\xi}{\Delta(\xi, l_0, m_0)} = \sum_{n=0,I}^{l_0} \frac{\pi(-1)^{r+1}k_n^{2r+1}e^{-k_n y}}{2C_n P(k_n; l_0)}, \quad r = 0, I, ..., l_0 - 1, \tag{2.68}$$

$$(c) \sum_{n=0,I}^{l_0} \frac{(-1)^r k_n^{2r}[\psi_n'(0)]^2}{C_n} + \frac{4}{\pi} \int_0^\infty \frac{\xi^{2r}[L'(\xi, y, ; l_0, m_0)]^2 d\xi}{\Delta(\xi, l_0, m_0)} = 0,$$

$$for \quad r = 0, I, ..., 2l_0 - 1, \tag{2.69}$$

where $\psi_n(y)$, $L(\xi, y; l_0, m_0)$ and C_n are the same as defined in Eqs. (2.61) and (2.62) with δ being the Dirac delta function.

Corollary 3 *The velocity potential $\phi(x,y)$ satisfying the governing equation (2.21), the bottom boundary condition in Eq. (2.22), the generalised condition in Eq. (2.56) on the mean structural boundary and the radiation condition as in (2.25) in water of finite depth is given by*

$$\phi(x,y) = \sum_{n=0,I}^{2l_0} A_n(x)\psi_n(y) + \sum_{n=1}^{\infty} A_n(x)\psi_n(y) \qquad (2.70)$$

where the eigenfunctions $\psi_n(y)$ are of the form given in Eq. (2.47) with p_n satisfying the generalised dispersion relation in Eq. (2.58). The eigenfunctions $\psi_n(y)$ in Eq. (2.70) satisfy the orthogonal mode-coupling relation

$$\left\langle \psi_m, \psi_n \right\rangle = C_n \delta_{mn} \quad for \quad m,n = 0,I,...,2l_0,1,2,..., \qquad (2.71)$$

where

$$\begin{aligned}
\left\langle \psi_m, \psi_n \right\rangle &= \int_0^h \psi_m(y)\psi_n(y)dy + \sum_{j=1}^{n_0} \frac{(-1)^j c_j}{Q(p_n; m_0)} \sum_{k=1}^{j} \psi_m^{2k-1}(0)\psi_n^{2j-(2k-1)}(0) \\
&+ \sum_{j=1}^{m_0} (-1)^{j+1} \frac{d_j p_n}{P(p_n; l_0)} \sum_{k=1}^{j} \psi_m^{2k-2}(0)\psi_n^{2j-2k}(0),
\end{aligned} \qquad (2.72)$$

$$C_n = \frac{\mathcal{G}'(k_n)}{2P(k_n; l_0)}, \quad for \quad n = 0,I,II,...,2l_0,1,2,..., \qquad (2.73)$$

with $\mathcal{G}(k)$, $P(k; l_0)$ and $Q(k; m_0)$ being the same as defined in Eq. (2.58). The unknown functions $A_n(x)$ are of the form as in Eq. (2.48) and are given by

$$A_n(x) = <\phi(x,y), \psi_n(y)>/C_n, \quad for \quad n = 0,I,II,1,2,... \qquad (2.74)$$

Lemma 9 *The orthogonal mode-coupling relation in Eq. (2.71) is equivalent to the relation given by*

$$\left\langle \psi_m, \psi_n \right\rangle = \int_0^h \left[\left(\psi_m''\psi_n' + \psi_m'\psi_n'' \right) + H_{mn} \left(\psi_n''\psi_m - \psi_n\psi_m'' \right) \right] dy, \qquad (2.75)$$

where

$$H_{mn} = \frac{Q(p_m; m_0)Q(p_n; m_0)}{Q(p_n; m_0)P(p_m; l_0) - Q(p_m; m_0)P(p_n; l_0)}. \qquad (2.76)$$

It may be noted that a similar orthogonal relation was given by [77] arising in acoustic structure interaction problems. A similar orthogonal relation to study wave interaction with floating structure was derived in [121]. Later, the equivalence of the orthogonal relations given in Eqs. (2.71) and (2.75) was established in [87].

Corollary 4 *The eigenfunctions $\psi_n(y)$ defined in Eq. (2.70) are linearly dependent.*

Proof 5 *The proof is a generalisation of Lemma 2 discussed in subsection 2.3.3 and details are deferred here.* □

After having an understanding about boundary value problems associated with the Laplace equation satisfying higher order boundary conditions in two dimensions. the full solution and associated results are discussed in three dimensions. Under the assumption of the linearised theory of water waves, the problem is considered in the three-dimensional Cartesian co-ordinate system with the x-z plane being horizontal and the y-axis considered positive in the vertically downward direction. The fluid domain is considered as an infinitely extended channel of finite/infinite width in water of finite and infinite depths. It is assumed that an ice sheet/large flexible plate is floating on the mean free surface. Along the z-axis, the plate is assumed to be of width b in the channel of finite width and is infinitely extended along the positive direction of the x-axis. Thus, the fluid domain occupies the region $0 < x < \infty$, $0 \leq y \leq h$ and $0 \leq z \leq b$ in the case of finite water depth ($0 < x < \infty$, $0 < y < \infty$ and $0 \leq z \leq b$ in the case of infinite water depth). The two side walls and the bottom of the channel are considered rigid. Thus, as discussed in the previous section and earlier in Section 1.4.3, the spatial velocity potential $\phi(x, y, z)$ satisfies the three-dimensional Laplace equation

$$\nabla^2 \phi = 0, \quad \text{in the fluid region,} \tag{2.77}$$

along with the bottom boundary conditions as in Eqs. (2.22) and (2.23). Assuming the channel is of finite width b, along the width of the channel, the side wall boundary conditions are assumed to satisfy

$$\phi = 0, \ z = 0, b, \quad 0 < x < \infty, \tag{2.78}$$

along with the infinity condition as given by

$$\phi, |\nabla \phi| \to 0 \quad \text{as} \quad z \to \infty. \tag{2.79}$$

On the structural boundary, the general type of higher order boundary condition is of the form

$$\left\{ \sum_{j=0}^{l_0} c_j \left(\frac{\partial^2}{\partial x^2} + \frac{\partial^2}{\partial z^2} \right)^j \frac{\partial}{\partial y} + \sum_{j=0}^{m_0} d_j \left(\frac{\partial^2}{\partial x^2} + \frac{\partial^2}{\partial z^2} \right)^j \right\} \phi = 0 \quad \text{on} \quad y = 0, \tag{2.80}$$

where c_j and d_j are known constants and l_0 and m_0 are positive integers with $m_0 \leq l_0$. Using Eq. (2.77), the boundary condition in Eq. (2.80) is rewritten as

$$(\mathcal{L} + \mathcal{M})\phi = 0, \quad y = 0, \ 0 \leq x < \infty, \ 0 \leq z \leq b, \tag{2.81}$$

where $\mathcal{L}$ and $\mathcal{M}$ are linear differential operators of the form

$$\mathcal{L} \equiv \sum_{j=0}^{l_0} (-1)^j c_j \frac{\partial^{2j+1}}{\partial y^{2j+1}}, \quad \mathcal{M} \equiv \sum_{j=0}^{m_0} (-1)^j d_j \frac{\partial^{2j}}{\partial y^{2j}}. \tag{2.82}$$

It may be noted that the higher order condition in Eq. (2.81) is the same as defined in Eq. (2.56) in the case of the two-dimensional problem. The far field radiation conditions as $x \to \infty$ are given by

$$
\phi(x,y,z) \sim
\begin{cases}
\displaystyle\sum_{m=1}^{\infty} R_{m0} e^{ip_{m0}x} \psi_{m0}(y,z), & \text{in finite water depth,} \\
\displaystyle\sum_{m=1}^{\infty} R_{m0} e^{ip_{m0}x} \psi_{m0}(y,z), & \text{in infinite water depth,}
\end{cases}
\tag{2.83}
$$

where

$$
\psi_{m0}(y,z) = \frac{\cosh k_0(h-y)}{\cosh k_0 h} \sin \frac{m\pi z}{b}, \quad p_{m0} = \sqrt{k_0^2 - (m\pi/b)^2}
$$

and k_0 satisfies the dispersion relation as in Eq. (2.58). In the case of three-dimensional problems, the results developed for the two-dimensional cases will be generalised using the appropriate boundary condition and additional edge conditions required to be prescribed along boundary on the z-axis. A few results related to the expansion formulae are discussed next in a precise manner.

The velocity potential $\phi(x,y,z)$ satisfying the governing equation (2.77) along with the boundary conditions (2.22) and (2.78), the higher order structural boundary condition (2.81) and the radiation condition (2.83) in the case of finite water depth is given by (proceeding in a similar manner as in [103])

$$
\phi(x,y,z) = \sum_{m=1}^{\infty} \left\{ \sum_{n=0,I}^{2l_0} A_{mn}(x)\psi_{mn}(y,z) + \sum_{n=1}^{\infty} A_{mn}(x)\psi_{mn}(y,z) \right\}, \tag{2.84}
$$

where the eigenfunctions $\psi_{mn}(y,z)$ are given by

$$
\psi_{mn}(y,z) =
\begin{cases}
\dfrac{\cosh k_n(h-y)}{\cosh k_n h} \sin \dfrac{m\pi z}{b}, & n = 0, I, ..., 2l_0, \\[3mm]
\dfrac{\cos k_n(h-y)}{\cos k_n h} \sin \dfrac{m\pi z}{b}, & n = 1, 2, 3, ...,
\end{cases}
\tag{2.85}
$$

with $p_{mn}^2 = k_n^2 - (m\pi/b)^2$ for $n = 0, I, ..., 2l_0, 1, 2, 3, ...,$ $m = 1, 2, 3, ...$ with k_n being the roots of the dispersion relation in k as in Eq. (2.58). The unknowns $A_{mn}(x)$ are of the form $A_{mn}(x) = \hat{A}_{mn} e^{ip_{mn}x}$ and are given by

$$
A_{mn}(x) = \frac{< \phi(x,y,z), \psi_{mn}(y,z) >}{C_n}, \tag{2.86}
$$

where $\psi_{mn}(y,z)$ satisfies the orthogonal mode-coupling relation given by

$$
< \psi_{mn}, \psi_{pq} > = \int_0^b \left[\int_0^h \psi_{mn}\psi_{pq}\,dy + \sum_{j=1}^{l_0} \frac{(-1)^j c_j}{Q(k_n;m_0)} \sum_{r=1}^{j} \psi_{mn}^{2r-1}\psi_{pq}^{2j-(2r-1)} \right]_{y=0}
$$

$$+\sum_{j=1}^{m_0}\frac{(-1)^{j+1}d_jk_n}{P(k_n;l_0)}\sum_{r=1}^{j}\psi_{mn}^{2r-2}(0,z)\psi_{pq}^{2j-2r}(0,z)\bigg]dz=\frac{b}{2}C_n\delta_{mp}\delta_{nq}, \quad (2.87)$$

with δ_{mp}, δ_{nq} being the Kroneckar delta and C_n being the same as in Eq. (2.73).

Lemma 10 *The eigenfunctions $\psi_{mn}(y,z)$ as defined in Eq. (2.85) are linearly dependent (as in [103]).*

Proof 6 *Proceeding in a similar manner as in Lemma 6 of Subsection 2.3.3 with the integral*

$$I=\int_C f(r)dr, \qquad (2.88)$$

where $f(r)=\dfrac{r\psi_m(r,y,z)}{\Gamma(r)}$, $\Gamma(r)=\sin b\sqrt{r^2-p_{mn}^2}$, $\qquad (2.89)$

$$\psi_m(r,y,z)=\frac{\cosh r(h-y)}{\cosh rh}\sin\frac{m\pi z}{b}, \quad p_{mn}^2=k_n^2-\left(\frac{m\pi}{b}\right)^2, \qquad (2.90)$$

with the contour C being the same as in Lemma 6, it can be easily derived that

$$\sum_{n=0,I}^{2p_0}\frac{k_n}{\Gamma'(k_n)}\psi_{mn}(y,z)+\sum_{n=1}^{\infty}\frac{k_n}{\Gamma'(k_n)}\psi_{mn}(y,z)=0. \qquad (2.91)$$

Assuming that k_n, $n=0,I,...,2p_0,II,1,2,3,...$ are the simple zeros of $\Gamma(r)=0$, $\Gamma'(k_n)\neq 0$, Eq. (2.91) concludes that the eigenfunctions are linearly dependent.

Theorem 7 *The velocity potential $\phi(x,y,z)$ satisfying the governing equation (2.77) along with the boundary conditions (2.22), (2.25) and (2.81) in the case of infinite water depth is given by*

$$\begin{aligned}\phi(x,y,z) &= \sum_{m=1}^{\infty}\sum_{n=0,I}^{2l_0}A_{mn}(x)\psi_{mn}(y,z)\\ &+ \frac{2}{\pi}\sum_{m=1}^{\infty}\int_0^{\infty}\frac{A_m(x,\xi)L(\xi,y;l_0,m_0)d\xi}{\Delta(\xi,l_0,m_0)}\sin\frac{m\pi z}{b},\end{aligned} \qquad (2.92)$$

where

$$\begin{aligned}A_{mn}(x) &= A_{mn}e^{ip_{mn}x}=\frac{2}{b\,C_n}\int_0^b\bigg[\int_0^{\infty}\phi(x,y,z)\psi_{mn}(y,z)dy\\ &+ \sum_{j=1}^{l_0}\frac{(-1)^{j+1}c_j}{Q(k_n;m_0)}\sum_{r=1}^{j}\phi^{2r-1}(x,0,z)\psi_{mn}^{2j-2r+1}(0,z)\\ &+ \sum_{j=1}^{m_0}\frac{(-1)^{j+1}d_jk_n}{P(k_n;l_0)}\sum_{r=1}^{j}\phi^{2r-2}(x,0,z)\psi_{mn}^{2j-2r}(0,z)\bigg]dz,\end{aligned}$$

$$A_m(x, \xi) = A_m e^{-\zeta_m x} = \frac{2}{b} \int_0^b \left[\int_0^\infty L(\xi, y; l_0, m_0) \phi(x, y, z) dy \right.$$

$$+ \sum_{j=1}^{l_0} \frac{(-1)^{j+1} c_j}{Q(i\xi; l_0)} \sum_{r=1}^{j} \xi^{2(j-r)+1} \phi^{2r-1}(x, 0, z)$$

$$+ \left. \sum_{j=1}^{m_0} \frac{(-1)^{j+1} d_j}{P(i\xi; l_0)} \sum_{r=1}^{j} \xi^{2(j-r)+1} \phi^{2r-2}(x, 0, z) \right] \sin \frac{m\pi z}{b} \, dz,$$

with $L(\xi, y; l_0, m_0)$ being the same as in Eq. (2.61), the eigenfunctions $\psi_{mn}(y, z)$ are given by

$$\psi_{mn}(y, z) = e^{-k_n y} \sin \frac{m\pi z}{b}, \qquad (2.93)$$

with k_n satisfying the following dispersion relation

$$\mathcal{F}(k) \equiv P(k; l_0) - Q(k; m_0) = 0 \qquad (2.94)$$

and p_{mn} being the same as in Eq. (2.90). It is assumed that the dispersion relation as in Eq. (2.94) has one positive real root k_0 which indicates the mode of the progressive wave and all the other roots are complex and are of the forms $\alpha_n \pm i\beta_n$, for $n = I, II, ..., 2l_0$. The details characteristic of the roots can be analysed knowing the unknown constants c_j, d_j as in the structural boundary (2.81). Further, C_n and $\Delta(\xi, l_0, m_0)$ are as in Eqs. (2.62) and (2.63), respectively.

Corollary 5 *The eigenfunctions described in Eq. (2.93) satisfy the following orthogonal mode-coupling relation*

$$< \psi_{mn}, \psi_{pq} > = \int_0^b \left[\int_0^\infty \psi_{mn}(y, z) \psi_{pq}(y, z) dy \right.$$

$$+ \sum_{j=1}^{l_0} \frac{(-1)^j c_j}{Q(k_n; m_0)} \sum_{r=1}^{j} \psi_{mn}^{2r-1}(0, z) \psi_{pq}^{2j-(2r-1)}(0, z)$$

$$+ \left. \sum_{j=1}^{m_0} \frac{(-1)^{j+1} d_j k_n}{P(k_n; l_0)} \sum_{r=1}^{j} \psi_{mn}^{2r-2}(0, z) \psi_{pq}^{2j-2r}(0, z) \right] dz$$

$$= \frac{b}{2} C_n \delta_{mp} \delta_{nq}. \qquad (2.95)$$

where δ_{mp}, δ_{nq} are Kronecker delta, and nonzero constant C_n is given in Eq. (2.62). The kernel $L(\xi, y; l_0, m_0)$ as in Eq. (2.61) and the eigenfunctions $\psi_{mn}(y, z)$ in Eq. (2.93) also satisfy the described orthogonal mode-coupling relation

$$< \psi_{mn}(y, z), L(\xi, y; l_0, m_0) \sin \frac{m\pi z}{b} >= 0. \qquad (2.96)$$

Theorem 8 *Given the coefficients* $A_{mn}(x)$ *and* $A_m(x, \xi)$ *as in equations (2.92) and (2.71), where* $\phi(x, y, z)$ *satisfies Laplace equation in the region* $0 < x < \infty$, $0 < y < \infty$, $0 \leq z \leq b$ *along with the boundary conditions (2.22), (2.25) and (2.81) and the eigenfunctions* $\psi_{mn}(y, z)$ *are defined by the equation (2.92), then the series*

$$\sum_{m=1}^{\infty} \sin \frac{m\pi z}{b} \left\{ \sum_{n=0,I}^{l_0} A_{mn}(x)e^{-k_n y} + \frac{2}{\pi} \int_0^\infty \frac{A(x, \xi)L(\xi, y; l_0, m_0)}{\Delta(\xi, l_0, m_0)} d\xi \right\} \quad (2.97)$$

converges to $\phi(x, y, z)$, *where* $\Delta(\xi, l_0, m_0)$ *is the same as in Theorem 7.*

Proof 7 *The proof follows from Theorem 5 and the identity given by (as in [31])*

$$\sum_{m=1}^{\infty} \sin \frac{m\pi z}{b} \sin \frac{m\pi p}{b} = \frac{b}{2}\delta(z - p), \quad \textit{for} \quad 0 < z, p < b. \quad (2.98)$$

$\square$

2.4 Flexural gravity wave maker problem

In this section, as with applications of the expansion formulae developed for two-dimensional wave structure interaction problems, wave motion generated by a vertical wavemaker for the flexural gravity waves is analysed in the cases of both infinite and finite water depths. The half plane $-\infty < x < \infty, 0 < y < \infty$ in the case of infinite depth and an infinite strip $-\infty < x < \infty, 0 < y < h$ in the case of finite water depth refer to the fluid domain associated with the physical problem. The motion considered here is two-dimensional and time-harmonic in nature with angular frequency ω and is due to the harmonically oscillating vertical plane wavemaker placed at $x = 0$ oscillating with velocity $Re\{U(y)e^{-i\omega t}\}$ with outgoing waves produced at far fields. The fluid surface is covered entirely by a floating ice sheet, which is under uniform compressive force N (as in Section 1.5). The spatial velocity potential $\phi(x, y)$ satisfies the Laplace equation (2.21) along with the bottom boundary conditions given in Eqs. (2.22) and (2.23) and the plate covered boundary condition as in Eq. (2.24). The boundary condition on the wavemaker is given by

$$\phi_x = U(y) \quad \text{on} \quad x = 0. \quad (2.99)$$

The radiation conditions are of the forms given in Eq. (2.25) with k_0 satisfying the dispersion relation in Eq. (2.26). For the uniqueness of the boundary value problem, depending on the physical problem, edge conditions are to be prescribed. In the context of the present problem, it is assumed that the sheer force and the bending moment are prescribed at the edge of the elastic plate, which yields

$$\begin{cases} D\phi_{xxxy} + \gamma Q\phi_{xy} \rightarrow \mu_1 & \text{as} \quad x \rightarrow 0^+, y = 0 \\ \phi_{xxy} \rightarrow \mu_2 & \text{as} \quad x \rightarrow 0^+, y = 0, \end{cases} \qquad (2.100)$$

where γ, μ_1 and μ_2 are known constants. As a direct application of Theorem 4, the velocity potential in the case of infinite water depth is given by

$$\phi(x,y) = -\sum_{j=0,I}^{II} A_j e^{i\epsilon_j p_j x - p_j y} + \frac{2}{\pi} \int_0^\infty \frac{\hat{U}(\xi) M(\xi,y) e^{-\xi x} d\xi}{\Delta(\xi)}, \qquad (2.101)$$

where $M(\xi,y), \Delta(\xi)$ are given by $M(\xi,y) = \xi(1 + Q\xi^2 + D\xi^4)\cos\xi y - K\sin\xi y$, $\Delta(\xi) = \xi^2(1 + Q\xi^2 + D\xi^4)^2 + K^2$, and $p_n(1 - Qp_n^2 + Dp_n^4) = K$ for $n = 0, I, II$. The dispersion relation has a positive real root p_0, which describes the progressive wave and two complex conjugate pairs (p_I, p_{II}) and (p_{III}, p_{IV}) with $p_{II} = \bar{p}_I$ and $p_{IV} = \bar{p}_{III}$. Here, it is assumed that the two roots p_I and p_{II} with positive real parts are lying in the first and fourth quadrants, respectively, and the other two roots are neglected because of the boundedness of the solution (see [14], [33]). Applying the boundary condition (2.99) on the wavemaker, and utilising the mode coupling relation (2.30), the unknown coefficients are obtained as

$$\hat{U}(\xi) = -\frac{1}{\xi}\int_0^\infty M(\xi,y)U(y)dy + (Q + D\xi^2)\alpha + D\beta, \qquad (2.102)$$

$$A_n = \frac{i\epsilon_n}{p_n C_n}\left[\int_0^\infty I_n(y)U(y)dy - \frac{p_n}{K}(Q + Dp_n^2)\alpha - \frac{Dp_n}{K}\beta\right], \qquad (2.103)$$

where $C_n = \dfrac{1 + 3Qp_n^2 + 5Dp_n^4}{2K}, \quad \epsilon_n = \begin{cases} -1 & \text{for} \quad n = 0, I, \\ 1 & \text{for} \quad n = II, \end{cases}$

$$I_n(y) = e^{-p_n y}, \quad n = 0, I, II,$$

with $\phi_{xy}(0^+, 0) = \alpha$, $\phi_{xyyy}(0^+, 0) = \beta$ and α, β are to be determined from the plate edge conditions as given in equation (2.100). Further, as a direct application of Theorem 3, velocity potential $\phi(x,y)$ in the case of finite depth is given by

$$\phi(x,y) = \sum_{j=0}^{II} A_j I_j(y) e^{ip_j \epsilon_j x} + \sum_{n=1}^\infty B_n I_n(y) e^{-p_n x}, \qquad (2.104)$$

where I_n is given by

$$I_n(y) = \begin{cases} \cosh p_n(h-y), & n = 0, I, II \\ \cos p_n(h-y), & n = 1, 2, ... \end{cases} \qquad (2.105)$$

and p_n satisfies the dispersion relations

$$K = \begin{cases} p_n(1 - Qp_n^2 + Dp_n^4)\tanh p_n h, & n = 0, I, II \\ -p_n(1 + Qp_n^2 + Dp_n^4)\tan p_n h, & n = 1, 2, ..., \end{cases} \qquad (2.106)$$

with $p_{II} = \bar{p}_I$, and p_I lies in the first quadrant. The complex roots are chosen to have positive real parts in order to satisfy the behavior of the solution at infinity. The unknown constants A_n for $n = 0, I, II$ are given by

$$A_n = \epsilon_n \mathcal{C}_n \left[\int_0^h U(Y) I_n(y) dy + \frac{D}{K} \{ \beta I'_n(0) + \alpha I'''_n(0) \} - \frac{Q}{K} \alpha I'_n(0) \right], \quad (2.107)$$

with $\mathcal{C}_n$ being the same as in Eq. (2.31). It is observed that all the unknowns are expressed in terms of the α and β which will be determined from the edge conditions (2.100).

2.5 Effect of compression on wave scattering by a crack

2.5.1 General introduction

As an application of the expansion formulae discussed in Section 2.3, the ice-coupled oblique wave propagation across an open crack in a uniformly compressed ice sheet in water of both infinite and finite depths is analysed in three dimensions in the linearised theory of water waves. Because of the symmetry in the geometry of the physical problem about the vertical plane along the crack, the associated BVP is divided into symmetric and antisymmetric parts and solved independently by using a Fourier-type expansion formulae and the associated mode-coupling relation developed in Section 2.3. Explicit expressions for the reflection and transmission coefficients are derived. The reflection and transmission coefficients are computed and analysed to understand the effect of crack and compression on the flexural gravity wave propagating below the ice sheet.

2.5.2 Mathematical formulation

The problem under consideration is three-dimensional in nature with the fluid occupying the half plane $-\infty < x, z < \infty$, $0 < y < \infty$ in the case of infinite depth and the infinite strip $-\infty < x, z < \infty$, $0 < y < h$ in the case of finite depth. A thin infinite ice sheet, which is under uniform compression N is floating on the upper surface $y = 0$, $-\infty < x, z < \infty$ as shown in Figure 2.1. The floating ice sheet of thickness d is modeled as two semi-infinite floating elastic plates separated by a long open crack at $x = 0$ as in Figure 2.1. With the assumption that the motion is simple harmonic in time and in the z direction, there exists a velocity potential $\Phi(x, y, z, t)$ of the form $\Phi(x, y, z, t) = Re\{\phi(x, y)e^{ilz - i\omega t}\}$, where ω is the angular frequency and l is the component of the wave number in the z direction. It is assumed that $y = \zeta(x, z, t)$ is the surface of the floating ice sheet with $\zeta(x, z, t)$ being the deflection of the floating ice sheet. Assuming that the plate deflection is symmetric along the

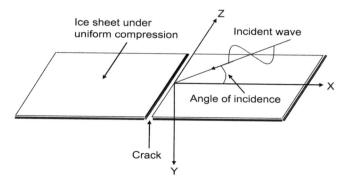

FIGURE 2.1
Schematic diagram of crack in floating ice sheet.

z-axis, the plate deflection is written in the form $\zeta(x,z,t) = Re\{\eta(x)e^{ilz-i\omega t}\}$. Thus, the spatial velocity potential $\phi(x,y)$ satisfies the PDE

$$(\nabla^2 - l^2)\phi = 0, \quad \text{in the fluid region.} \tag{2.108}$$

The linearised mean ice-covered surface boundary condition at $y = 0$ in the presence of uniform compressive force N is given by (see Eq. (1.108) of Chapter 1 and Eq. (2.24) of Chapter 2)

$$D\frac{\partial^5 \phi}{\partial y^5} - Q\frac{\partial^3 \phi}{\partial y^3} + \frac{\partial \phi}{\partial y} + K\phi = 0, \tag{2.109}$$

where the constant coefficients are the same as described in Subsection 1.5.3. The bottom boundary condition is given by

$$\phi, |\nabla\phi| \to 0, \quad \text{as} \quad y \to \infty, \quad \text{in water of infinite depth,}$$
$$\frac{\partial \phi}{\partial y} = 0, \quad \text{on} \quad y = h, \quad \text{in water of finite depth.} \tag{2.110}$$

Continuity of velocity and pressure across the vertical interface (often called matching condition) at $x = 0$ is given by

$$\frac{\partial\phi(0^-, y)}{\partial y} = \frac{\partial\phi(0^+, y)}{\partial y}, \quad \phi(0^-, y) = \phi(0^+, y). \tag{2.111}$$

Assuming that near the crack, the ice sheets behave like free edges, the edge conditions at $x = 0^\pm$, $y = 0$ are prescribed as

$$B\phi \equiv \left[\left\{\frac{\partial^2}{\partial x^2} - \nu l^2\right\}\frac{\partial\phi}{\partial y}\right]_{y=0} \to 0, \quad \text{as} \quad x \to 0^\pm, \tag{2.112}$$

$$S\phi \equiv \left[D\frac{\partial}{\partial x}\left\{\frac{\partial^2}{\partial x^2} - \nu_1 l^2\right\}\frac{\partial\phi}{\partial y} + Q\frac{\partial^2\phi}{\partial x\partial y}\right]_{y=0} \to 0, \quad \text{as} \quad x \to 0^\pm, \tag{2.113}$$

where $\nu_1 = 2 - \nu$. The far field radiation condition in the case of infinite water depth is given by

$$\phi \sim \begin{cases} (e^{-i\gamma_0 x} + R_0 e^{i\gamma_0 x})e^{-p_0 y}, x \to \infty, \\ T_0 e^{-i\gamma_0 x} e^{-p_0 y}, x \to -\infty \end{cases} \tag{2.114}$$

and
$$\phi \sim \begin{cases} (e^{-i\gamma_0 x} + R_0 e^{i\gamma_0 x})\dfrac{\cosh p_0(h-y)}{\cosh p_0 h}, x \to \infty, \\ T_0 e^{-i\gamma_0 x}\dfrac{\cosh p_0(h-y)}{\cosh p_0 h}, x \to -\infty, \end{cases} \tag{2.115}$$

in the case of finite water depth, where R_0 and T_0 are the complex amplitudes of the reflected and transmitted waves, respectively, and p_0 is the positive real root of the dispersion relation in p for the ice covered region as given by

$$K = \begin{cases} (Dp^4 - Qp^2 + 1)p & \text{in the case of infinite depth,} \\ (Dp^4 - Qp^2 + 1)p \tanh ph & \text{in the case of finite depth,} \end{cases} \tag{2.116}$$

along with $l = p_0 \sin\theta$, $\gamma_0 = p_0 \cos\theta$. The amplitude of the incident wave is taken as unity for numerical convenience.

2.5.3   Method of solution

It is clear from Figure 2.1 that the physical problem is symmetric about the plane $x = 0$ in the cases of both infinite and finite depths. Thus, exploiting the geometrical symmetry of the physical problem, the physical problems defined in the half plane and the infinite strip are reduced to problems in quarter plane and semi-infinite strip, respectively, and then the expansion formulae and the associated mode-coupling relations, as developed in Section 2.3, are applied to obtain the full solution. Thus, the total velocity potential $\phi(x, y)$ is written as the sum of symmetric and antisymmetric potential functions as

$$\phi = \frac{1}{2}(\phi_s + \phi_a), \tag{2.117}$$

where ϕ_s and ϕ_a are even and odd functions respectively about $x = 0$. Using the properties of even and odd functions, we have

$$\phi_s(-x, y) = \phi(x, y) + \phi(-x, y) = \phi_s(x, y),$$
$$\phi_a(-x, y) = -\phi(x, y) + \phi(-x, y) = -\phi_a(x, y). \tag{2.118}$$

Utilising the relations in Eq. (2.118), the boundary value problems associated with the physical problems defined in water of infinite and finite depths are reduced to two independent BVPs in the quarter plane $x > 0, y > 0$ and semi-infinite strip $x > 0, 0 < y < h$, respectively. With the help of the symmetric and anti-symmetric properties, the continuity of velocity and pressure across the crack along $x = 0$ yields

$$\frac{\partial \phi_s(0, y)}{\partial x} = \phi_a(0, y) = 0, \quad 0 \le y < \infty. \tag{2.119}$$

Further, ϕ_s and ϕ_a are divided into incident and scattered potentials and written as

$$\left.\begin{array}{l} \phi_s(x,y) = \phi_{0s}(x,y) + \psi_s(x,y), \\ \phi_a(x,y) = \phi_{0a}(x,y) + \psi_a(x,y), \end{array}\right\} \quad \text{for} \quad x \geq 0, \qquad (2.120)$$

where $\phi_{0s}(x,y) = (e^{-i\gamma_0 x} + e^{i\gamma_0 x})e^{-p_0 y}$ and $\phi_{0a}(x,y) = (e^{-i\gamma_0 x} - e^{i\gamma_0 x})e^{-p_0 y}$ are, respectively, the symmetric and anti-symmetric incident standing waves in the case of infinite depth and $e^{-p_0 y}$ will be replaced by $(\cosh p_0(h - y)/\cosh p_0 h)$ in the case of finite depth. The functions ψ_s and ψ_a represent the symmetric and anti-symmetric scattered potentials and satisfy the governing Eq. (2.108) along with the ice-covered boundary condition (2.109) and the bottom boundary condition in Eq. (2.110) as appropriate for infinite or finite water depth. However, the far field behavior is of the form

$$\psi_{s,a}(x,y) \sim R_{s,a}e^{i\gamma_0 x - p_0 y}, \text{ as } x \to \infty, \qquad \text{for infinite depth, } (2.121)$$

$$\psi_{s,a}(x,y) \sim R_{s,a}e^{i\gamma_0 x}\frac{\cosh p_0(h - y)}{\cosh p_0 h}, \text{ as } x \to \infty, \text{ for finite depth, } (2.122)$$

where $R_{s,a}$ are unknown constants to be determined. Comparing the radiation conditions in Eqs. (2.114), (2.115), (2.121) and (2.122) leads to the following relations related to the amplitudes R_0 and T_0 of the reflected and transmitted waves as given by

$$R_0 = (R_s + R_a)/2 \quad \text{and} \quad T_0 = 1 + (R_s - R_a)/2. \qquad (2.123)$$

In addition, using the definitions of ϕ_s and ϕ_a from Eq. (2.118), the edge conditions in Eqs. (2.112) and (2.113) are modified to give

$$B\psi_s = -2(\gamma_0^2 + \nu l^2)p_0, \ B\psi_a = \alpha_a - (1 - \nu)l^2\beta_a = 0,$$
$$S\psi_s = \alpha_s - (1 - \nu_1)l^2\beta_s = 0, \ S\psi_a = 2i\{(\gamma_0^2 + \nu_1 l^2)D - Q\}\gamma_0 p_0, \qquad (2.124)$$

in the case of infinite depth and

$$B\psi_s = -2(\gamma_0^2 + \nu l^2)p_0\tanh p_0 h, \ B\psi_a = \alpha_a - (1 - \nu)l^2\beta_a = 0,$$
$$S\psi_s = \alpha_s - (1 - \nu_1)l^2\beta_s = 0, \ S\psi_a = 2i\{D(\gamma_0^2 + \nu_1 l^2) - Q\}\gamma_0 p_0\tanh p_0 h, \qquad (2.125)$$

in the case of finite depth, with

$$\alpha_s = \frac{\partial^4\psi_a(0,0)}{\partial x\partial y^3}, \quad \alpha_a = \frac{\partial^3\psi_a(0,0)}{\partial y^3}, \quad \beta_s = \frac{\partial^2\psi_s(0,0)}{\partial x\partial y}, \quad \beta_a = \frac{\partial\psi_s(0,0)}{\partial y}.$$

Next, we proceed to solve for the symmetric and anti-symmetric potentials independently in the cases of both infinite and finite depths. First we consider the case of infinite depth.

Infinite depth: Using the expansion formula as in Section 2.3, the symmetric potential ψ_s is derived as

$$\psi_s(x,y) = \sum_{j=0}^{II} A_j e^{i\epsilon\gamma_j x - p_j y} + \frac{2}{\pi}\int_0^\infty \frac{M(\xi,y)A(\xi)e^{-\zeta x}}{\Delta(\xi)}d\xi, \qquad (2.126)$$

where

$$A_n = \frac{i\epsilon_n p_n}{LC_n\gamma_n}\{D(\gamma_n^2 + \nu l^2) - Q\}\beta_s, \qquad \epsilon_n = \begin{cases} 1, & n = 0, 1, \\ -1, & n = 2, \end{cases}$$

$$A(\xi) = -\frac{\xi}{\zeta}\{D(\zeta^2 - \nu l^2) + Q\}\beta_s,$$

$$\beta_s = -\frac{2\{(\gamma_0^2 + \nu l^2)D - Q\}p_0}{P_s},$$

$$P_s = \frac{i}{L}\sum_{n=0}^{2}\frac{\epsilon_n p_n^2\{D(\gamma_n^2 + \nu l^2) - Q\}(\gamma_n^2 + \nu l^2)}{C_n\gamma_n}$$
$$+ \frac{2L}{\pi}\int_0^\infty \frac{\{D(\zeta^2 - \nu l^2) + Q\}(\zeta^2 - \nu l^2)\xi^2}{\zeta\Delta(\xi)}d\xi,$$

and $M(\xi, y)$ and $\Delta(\xi)$ are same as given in Eq. (2.101).

Proceeding in a similar manner as in the case of symmetric potential, the antisymmetric potential ψ_a is derived as

$$\psi_a(x, y) = \sum_{j=0}^{II} B_j e^{i\epsilon_j\gamma_j x - p_j y} + \frac{2}{\pi}\int_0^\infty \frac{M(\xi, y)B(\xi)e^{-\zeta x}d\xi}{\Delta(\xi)}, \qquad (2.127)$$

where

$$B_n = \frac{p_n}{LC_n}\{Q - D(\gamma_n^2 + \nu_1 l^2)\}\beta_a, \quad n = 0, 1, 2,$$
$$B(\xi) = \{Q + D(\zeta^2 - \nu_1 l^2)\}\xi\beta_a,$$
$$\beta_a = 2i\gamma_0\{(\gamma_0^2 + \nu_1 l^2)D - Q\}p_0/P_a,$$
$$P_a = \frac{i}{L}\sum_{n=0}^{2}\frac{\epsilon_n\{Q - D(\gamma_n^2 + \nu_1 l^2)\}(\gamma_n^2 + \nu_1 l^2)\gamma_n p_n^2}{C_n}$$
$$+ \frac{2L}{\pi}\int_0^\infty \frac{\{Q + D(\zeta^2 - \nu_1 l^2)\}(\zeta^2 - \nu_1 l^2)\xi^2\zeta}{\Delta(\xi)}d\xi,$$

where ϵ_n and C_n are as defined in Eq. (2.103). The integrals for β_s and β_a in Eqs. (2.126) and (2.127), respectively, are integrable and can be easily evaluated numerically. Once the unknown constants A_0 and B_0 are obtained, the complex amplitudes of the reflection and transmission coefficients, R_0 and T_0, respectively, are evaluated from relation (2.123) by replacing R_s and R_a by A_0 and B_0. The reflection and transmission coefficients are defined as $K_r = |R_0|$ and $K_t = |T_0|$, respectively. Using Green's integral theorem, the energy relation is obtained as $K_r^2 + K_t^2 = 1$, which is the same as the energy relation for the free surface gravity waves.

Finite depth: In the case of finite depth, proceeding in a similar manner,

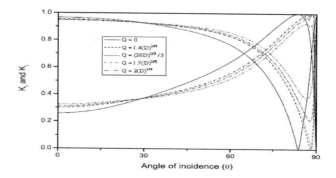

FIGURE 2.2
Variation of K_r and K_t versus angle of incidence θ for various values of the compressive force Q with $\lambda = 40$m.

the expansions for the symmetric and antisymmetric potentials are given by

$$\psi_s(x,y) = A_0 e^{i\gamma_0 x} f_0(y) + A_I e^{i\gamma_I x} f_I(y)$$
$$+ A_{II} e^{-i\gamma_{II} x} f_{II}(y) + \sum_{n=1}^{\infty} A_n e^{-\gamma_n x} f_n(y), \qquad (2.128)$$

$$\psi_a(x,y) = B_0 e^{i\gamma_0 x} f_0(y) + B_I e^{i\gamma_I x} f_I(y)$$
$$+ B_{II} e^{-i\gamma_{II} x} f_{II}(y) + \sum_{n=1}^{\infty} B_n e^{-\gamma_n x} f_n(y), \qquad (2.129)$$

where p_0 is the real and p_n, $n = I, II$ are the complex roots of the dispersion relation (2.116), $\gamma_n = \sqrt{p_n^2 - l^2}$ and

$$f_n(y) = \frac{\cosh p_n(h-y)}{\cosh p_n h}, \quad n = 0, I, II, \qquad (2.130)$$

and $p_n = ip_n$ for $n = 1, 2, ...$ In this case also, by the direct application of the expansion formulae as discussed in Section 2.3, the unknown coefficients A_n and B_n are obtained as

$$A_n = \frac{i\epsilon_n p_n}{LC_n \gamma_n}\{(\gamma_n^2 + \nu l^2)D - Q\}\beta_s \tanh p_n h, \quad n = 0, I, II,$$

$$A_n = \frac{p_n}{LC_n \gamma_n}\{(\gamma_n^2 - \nu l^2)D + Q\}\beta_s \tan p_n h, \quad n = 1, 2, ..., \qquad (2.131)$$

$$B_n = -\frac{p_n}{LC_n}\{(\gamma_n^2 + \nu_1 l^2)D - Q\}\beta_a \tanh p_n h, \quad n = 0, I, II,$$

$$B_n = -\frac{p_n}{LC_n}\{(\gamma_n^2 - \nu_1 l^2)D + Q\}\beta_a \tan p_n h, \quad n = 1, 2, ..., \qquad (2.132)$$

$$\text{where } \beta_s = -\frac{2\{(\gamma_0^2 + \nu l^2)D - Q\}p_0 \tanh p_0 h}{D_s},$$

$$\beta_a = \frac{2i\gamma_0\{(\gamma_0^2 + \nu_1 l^2)D - Q\}p_0 \tanh p_0 h}{D_a},$$

$$D_s = \frac{i}{L}\sum_{n=0,I}^{II} \frac{\epsilon_n p_n^2\{(\gamma_n^2 + \nu l^2)D - Q\}}{C_n\gamma_n}(\gamma_n^2 + \nu l^2)\tanh^2 p_n h$$

$$+ \frac{1}{L}\sum_{n=1}^{\infty} \frac{p_n^2\{(\gamma_n^2 - \nu l^2)D + Q\}}{C_n\gamma_n}(\gamma_n^2 - \nu l^2)\tan^2 p_n h,$$

$$D_a = -\frac{i}{L}\sum_{n=0,I}^{II} \frac{\epsilon_n\gamma_n p_n^2\{(\gamma_n^2 + \nu_1 l^2)D - Q\}}{C_n}(\gamma_n^2 + \nu_1 l^2)\tanh^2 p_n h$$

$$+ \frac{1}{L}\sum_{n=1}^{\infty} \frac{\gamma_n p_n^2\{(\gamma_n^2 - \nu_1 l^2)D + Q\}}{C_n}(\gamma_n^2 - \nu_1 l^2)\tan^2 p_n h,$$

with $p_n = ip_n$ for $n = 1, 2, ...,$ ϵ_n as defined in Eq. (2.103) and C_n as defined in Eq. (2.31).

2.5.4   Results and discussions

Simple numerical computations are carried out from the solutions to analyse the behaviour of reflection and transmission coefficients in the case of infinite depth. The results in the case of finite depth are expected to be similar in nature to the results of infinite depth. The effects of uniform compressive force Q, angle of incidence θ and thickness of the ice sheet d on various important parameters are analysed. The numerical values of the physical parameters, which are fixed throughout the computation, are $\rho = 1025$kg-m^{-3}, $\rho_p = 922.5$kg-m^{-3}, $\nu = 0.3$, $g = 9.81$m-sec^{-2}, $d = 1$m and $E = 5$GPa.

In Figure 2.2, the variation of transmission coefficient K_t versus the angle of incidence θ is plotted for various values of ice thickness d. For a particular ice thickness, as θ increases, K_r decreases continuously and after attaining the minimum value it increases sharply and becomes exactly one at $\theta = 90°$. For relatively thick ice sheet, the reflection coefficient is large for most incident wave angles. The reverse phenomenon occurs in the case of K_t. It may be noted that because of the existence of compressive force, the minimum in the reflection coefficients is nonzero for an intermediate value of the wave period unlike the case of zero compressive force as analysed in [33].

In Figure 2.3 and Figure 2.4, variation of the reflection and transmission coefficients K_r and K_t versus time period T for different values of compressive force Q are plotted. It is observed that there are certain values of the time period beyond which the ice compression does not affect the reflection and transmission coeffcients K_r and K_t. It is further noticed that as the time

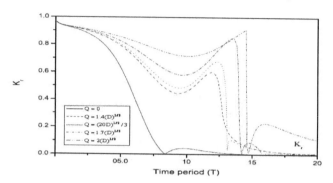

FIGURE 2.3

Variation of K_r versus time period T for various values of the compressive force Q with $\theta = 30°$.

period increases, the reflection coefficient decreases and the transmission coefficient increases. After attaining a minimum, the reflection coefficient increases for a certain interval of the time period and then again decreases, and the reverse pattern is observed in the case of the transmission coefficient. However, for nonzero compressive force, the optimum value in the reflection coefficient increases as Q increases. It is further noticed that K_r reaches the minimum value for the certain time period at which the phase velocity also attains the minimum. As discussed in Section 1.5, these extreme values are related to the critical limiting values of the compressive force on a floating ice sheet. A detailed analysis to this effect is in [59].

2.6   Expansion formulae in double layer fluid

While dealing with boundary value problems associated with the Laplace equation, apart from boundary conditions at the external boundaries, often singularities at the interface of two boundaries bring additional boundary conditions making the problem more complex. One of these problems is the wave motion in a two-layer fluid having free surface and the interface. A more general mathematical problem of this nature is the problem associated with the wave motion in the presence of floating and submerged elastic plates in the presence of compressive forces in a two-layer fluid of different densities ρ_1 and ρ_2 with $\rho_1 < \rho_2$. It is assumed that the floating plate is at the mean free surface and the submerged plate is at the mean interface. As a result, the boundary conditions at both the free surface and the interface are of the 5th order in nature. The mathematical problem is similar to the one discussed in

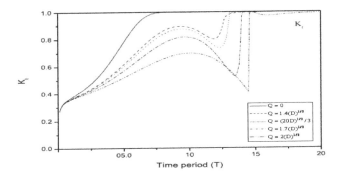

FIGURE 2.4
Variation of K_t versus time period T for various values of the compressive force Q with $\theta = 30°$.

Subsection 1.4.3. Unlike the case of the expansion formulae discussed in the single layer fluid, in this case, the complexity increases due to the presence of the submerged plate at the interface of the fluid domain. The general forms of the expansion formulae in water of finite and infinite depths are discussed and the expansion formulae are further generalised to deal with boundary value problems satisfying higher order boundary conditions of higher order at the free surface. Results associated with simpler physical problems are obtained as particular cases from this general expansion formulae. For specific physical problems, depending on the nature of the problem, appropriate edge conditions will be prescribed to find the unknowns associated with the general expansion formulae. Further, several characteristics of the eigen-system associated with the velocity potential are discussed. The utility of the expansion formulae is illustrated by analysing specific physical problems.

2.6.1 Wave structure interaction problems in two-layer fluid

Under the assumption of the linearised theory of water waves and small amplitude structural response, the problem is considered in the two-dimensional Cartesian coordinate system with the x-axis being in the horizontal direction and the y-axis being in the vertically downward positive direction. The problems are discussed in both cases of water of finite and infinite depths. An infinitely extended thin elastic plate is floating at the mean free surface $y = 0$ in an infinitely extended fluid which is assumed to be inviscid and incompressible and motion is irrotational. Another infinitely extended submerged flexible plate is kept horizontally at $y = h$ in the fluid domain. It is assumed that a fluid of density ρ_1 occupies the region $0 < y < h$, $0 < x < \infty$. On the other hand, the fluid of density ρ_2 occupies the region $h < y < H$, $0 < x < \infty$ in the case of finite water depth and $h < y < \infty$, $0 < x < \infty$ in the case of infinite water depth. Thus, as discussed in Subsection 1.4.3, the spatial velocity potential $\phi(x, y)$ satisfies the two-dimensional Laplace equation as given by

$$\nabla^2\phi = 0, \quad \text{in the fluid domain.} \tag{2.133}$$

The bottom boundary condition is given by

$$\frac{\partial\phi}{\partial y} = 0 \quad \text{at} \quad y = H, \quad \text{in the case of finite depth} \tag{2.134}$$

and

$$\phi, |\nabla\phi| \to 0 \quad \text{as} \quad y \to \infty, \quad \text{in the case of infinite depth.} \tag{2.135}$$

The boundary condition at the mean free surface at $y = 0$ on the floating flexible plate is given by

$$D_1\frac{\partial^5\phi}{\partial y^5} - Q_1\frac{\partial^3\phi}{\partial y^3} + \frac{\partial\phi}{\partial y} + K\phi = 0 \text{ on } y = 0, \ 0 \leq x < \infty. \tag{2.136}$$

The boundary conditions at the mean interface on the flexible submerged plate surface for $0 < x < \infty$ are given by

$$\frac{\partial\phi}{\partial y}\bigg|_{y=h+} = \frac{\partial\phi}{\partial y}\bigg|_{y=h-} \tag{2.137}$$

and

$$\left(D_2\frac{\partial^5\phi}{\partial y^5} - Q_2\frac{\partial^3\phi}{\partial y^3} + \frac{\partial\phi}{\partial y} + K\phi\right)\bigg|_{y=h+} = s\left(\frac{\partial\phi}{\partial y} + K\phi\right)\bigg|_{y=h-}. \tag{2.138}$$

In Eqs. (2.136) and (2.138), D_1, D_2, Q_1, Q_2, K and s are known positive constants as in Subsection 1.4.4.

2.6.2   Expansion formulae in infinite water depth

Theorem 9 *The velocity potential $\phi(x, y)$ satisfying Eq. (2.133) along with the boundary conditions Eqs. (2.135)–(2.138) is given by*

$$\phi(x, y) = \sum_{n=I}^{X} A_n F_n(k_n, y)e^{ik_n x} + \int_0^\infty A(\xi)L(\xi, y)e^{-\xi x}d\xi, \quad x > 0, \tag{2.139}$$

where

$$F_n(k_n, y) = \begin{cases} \dfrac{iL_1(ik_n, y)}{KD(ik_n, h)}, & 0 < y < h, \\ e^{-k_n(y-h)}, & h < y < \infty, \end{cases}$$

$$L(\xi, y) = \begin{cases} L_1(\xi, y) & for \quad 0 < y < h, \\ L_2(\xi, y) & for \quad h < y < \infty, \end{cases}$$

$$L_1(\xi, y) = K\{T_1(\xi; 2)\xi \cos \xi y - K \sin \xi y\},$$

$$L_2(\xi, y) = L_1(\xi, y) - W(\xi) \cos \xi(y - h),$$

$$W(\xi) = (1 - s)L_1(\xi, h) - \xi\{T_2(\xi; 2) - s\}D(\xi, h),$$

$$D(\xi, h) = T_1(\xi; 2)\xi \sin \xi h + K \cos \xi h.$$

$$T_j(\xi; 2) = D_j\xi^4 + Q_j\xi^2 + 1, \quad j = 1, 2,$$

with k_n satisfying the dispersion relation

$$G(k_n) \equiv K - \frac{k_n T_1(ik_n; 2)[K(s \coth k_n h + 1) - k_n\{T_2(ik_n; 2) - s\}]}{K(s + \coth k_n h) - k_n\{T_2(ik_n; 2) - s\} \coth k_n h} = 0. \tag{2.140}$$

In Eq. (2.139), keeping the realistic nature of physical problems in mind, it is assumed that the dispersion relation in Eq. (2.140) has two distinct real positive roots k_I and k_{II} and eight complex roots $k_{III}, k_{IV}, ..., k_X$ of the forms $-a \pm ib$ and $-c \pm id$. Thus, the boundedness criteria of the velocity potential $\phi(x, y)$ at the far field yields $A_{VII} = ... = A_X = 0$. Further, the eigenfunctions F_n and $L(\xi, y)$ satisfy the orthogonal mode-coupling relation given by

$$\langle F_m, F_n \rangle = s\langle F_m, F_n \rangle_1 + \langle F_m, F_n \rangle_2 = C_n \delta mn, \quad m, n = I, ..., VI, \tag{2.141}$$

with

$$\langle F_m, F_n \rangle_1 = \int_0^h F_m F_n dy + \left\{ \frac{D_1}{K}(F'_m F'''_n + F'''_m F'_n) - \frac{N_1}{K} F'_m F'_n \right\}\Big|_{y=0},$$

$$\langle F_m, F_n \rangle_2 = \lim_{\epsilon \to 0} \int_h^\infty e^{-\epsilon y} F_m F_n dy + \left\{ \frac{D_2}{K}(F'_m F'''_n + F'''_m F'_n) - \frac{N_2}{K} F'_m F'_n \right\}\Big|_{y=h},$$

where C_n is given by

$$C_n = \frac{-D(ik_n, h) \sinh k_n h [\psi'_n(0)]^2 G'(k_n)}{2K^2 k_n^2}, \tag{2.142}$$

with $G(k_n)$ as in (Eq. (2.40)). Again, it can be derived that

$$\langle F_n, L(\xi, y) \rangle = 0 \quad for \ all \ \xi > 0 \ and \ n = I, ..., VI. \tag{2.143}$$

Using the above-mentioned orthogonal mode-coupling relation, the unknowns $A_n, n = I, ..., VI, A(\xi)$ associated with the expansion formulae are obtained as

$$A_n e^{ik_n x} = \frac{\langle \phi(x, y), F_n(k_n, y) \rangle}{C_n}, \quad A(\xi)e^{-\xi x} = \frac{2\langle \phi(x, y), L(\xi, y) \rangle}{\pi \Delta(\xi)}, \tag{2.144}$$

with $\Delta(\xi) = W^2(\xi) - 2KW(\xi)L_1(\xi, h) + \{\xi^2 T_1^2(\xi; 2) + K^2\}K^2$. *Thus, the expansion formula for the velocity potential in the case of infinite water depth is derived completely.*

Corollary 6 *The equivalent form of the eigenfunctions* $F_n(k_n, y)$ *as in Eq. (2.139) is given by*

$$F_n(k_n, y) = \begin{cases} f_n(k_n, y) & for \quad 0 < y < h, \\ -Kse^{-k_n(y-h)} & for \quad h < y < \infty, \end{cases} \quad n = I, II; \quad (2.145)$$

where $f_n(k_n, y) = \left[\{T_2(ik_n; 2) - s\}k_n - K\right]L_1(ik_n, y)/L_1(ik_n, h)$. *This form of* $F_n(k_n, y)$ *is derived using the dispersion relation in Eq. (2.140) by using the alternate form of the dispersion relation as given by*

$$\frac{\{T_2(ik_n; 2) - s\}k_n - K}{iL_1(ik_n, h)} = \frac{s}{\mathcal{D}(ik_n, h)}. \quad (2.146)$$

2.6.3   Expansion formulae in finite water depth

The velocity potential $\phi(x, y)$, satisfying Eq. (2.133) along with the boundary conditions Eqs. (2.134), (2.136)–(2.138), is given by

$$\phi(x, y) = \sum_{n=I,...,X,1}^{\infty} B_n \psi_n(k_n, y) e^{ik_n x} \quad for \quad x > 0, \quad (2.147)$$

where the eigenfunctions ψ_n are obtained as

$$\psi_n(k_n, y) = \begin{cases} -\dfrac{i \tanh k_n (H - h) L_1(ik_n, y)}{K\mathcal{D}(ik_n, h)}, & 0 < y < h, \\ \dfrac{\cosh k_n (H - y)}{\cosh k_n (H - h)}, & h < y < H. \end{cases} \quad (2.148)$$

The eigenvalues $k_n, n = I, II, ..., IX, X$ satisfy the dispersion relation in k (see Eq. (1.118) of Chapter 1) as given by

$$\mathcal{G}(k) \equiv K - \frac{kT_1(ik; 2)}{\mu} = 0, \quad (2.149)$$

where

$$\mu = \frac{K\{s + \coth kh \coth k(H - h)\} - \{T_2(ik; 2) - s\}k \coth kh}{K\{s \coth kh + \coth k(H - h)\} - k\{T_2(ik; 2) - s\}}.$$

Keeping in mind that such boundary value problems are of physical interest, it is assumed that the dispersion relation in Eq. (2.149) has two distinct positive real roots k_n, $n = I, II$, eight complex roots k_n, $n = III, ..., X$ of the form $-a \pm ib$ and $-c \pm id$ and an infinite number of purely imaginary roots k_n, $n = 1, 2, ...$ of the form $k_n = ik_n$. However, a brief discussion about the behavior of wave numbers associated with progressive waves in the case of deep and shallow water waves is provided in Subsection 1.4.4. Assuming the boundedness property of the velocity potential, the eigenfunctions associated with the eigenvalues $k_{VII}, ..., k_X$ will not contribute to the expansion formula and thus $B_{VII} = ... = B_X = 0$ in Eq. (2.147). Further, the eigenfunctions ψ_n in Eq. (2.148) satisfy the orthogonal mode-coupling relation as given by

$$\langle \psi_m, \psi_n \rangle = s \langle \psi_m, \psi_n \rangle_3 + \langle \psi_m, \psi_n \rangle_4 = C_n \delta_{mn}, \quad m, n = I, ... VI, 1, 2, ...,$$
(2.150)

where

$$\langle \psi_m, \psi_n \rangle_3 = \int_0^h \psi_m \psi_n dy + \left\{ \frac{D_1}{K} (\psi'_m \psi'''_n + \psi'''_m \psi'_n) - \frac{Q_1}{K} \psi'_m \psi'_n \right\} \Bigg|_{y=0},$$

$$\langle \psi_m, \psi_n \rangle_4 = \int_h^H \psi_m(y) \psi_n(y) dy + \left\{ \frac{D_2}{K} (\psi'_m \psi'''_n + \psi'''_m \psi'_n) - \frac{Q_2}{K} \psi'_m \psi'_n \right\} \Bigg|_{y=h},$$

$$C_n = \frac{-\mathcal{D}(ik_n, h) \sinh k_n h [\psi'_n(0)]^2 \mathcal{G}'(k_n)}{2K^2 k_n^2},$$
(2.151)

which has the alternate form in terms of $\psi'_n(h)$ as given by

$$C_n = \frac{-\sinh k_n h [\psi'_n(h)]^2 \mathcal{G}'(k_n)}{2k_n^2 \mathcal{D}(ik_n, h)},$$
(2.152)

where $\mathcal{G}(k_n)$ is the same as defined in Eq. (2.149) and $\mathcal{D}(k_n, h)$ is as defined in Eq. (2.139). Using the orthogonal mode-coupling relation in Eq. (2.150), the unknown coefficients are obtained as

$$B_n = \frac{\langle \phi(x, y), \psi_n(k_n, y) \rangle}{C_n e^{ik_n x}}, \quad n = I, ..., VI, n = 1, 2, ...$$
(2.153)

This completes the derivation of the expansion formula and the associated mode-coupling relation in the case of finite water depth.

Lemma 11 *The eigenfunctions $\psi'_n(y)$ satisfy the following two identities as given by*

$$\sum_{n=I,...,VI,1}^{\infty} \frac{[\psi'_n(0)]^2}{C_n} = 1 \quad and \quad \sum_{n=I,...,VI,1}^{\infty} \frac{[\psi'_n(h)]^2}{C_n} = 1.$$
(2.154)

Proof 8 *Considering the integral*

$$I_1 = \frac{1}{2\pi i} \int_c \left[\frac{K^2 \alpha^2}{\mathcal{G}(\alpha) \sinh \alpha h \{\alpha T_1(i\alpha; 2) \sinh \alpha h - K \cosh \alpha h\}} - \frac{1}{\alpha} \right] d\alpha,$$
(2.155)

where the contour c is a circle, the centre being at origin of radius R, and using Jordan's lemma (proceeding in a similar manner as in [33]), it can be easily derived that

$$I_1 = 2 \sum_{n=I,...,VI,1}^{\infty} \frac{K^2 k_n^2}{\mathcal{G}'(k_n)\sinh k_n h \{k_n T_1(ik_n;2)\sinh k_n h - K\cosh k_n h\}} - 1$$

$$= \sum_{n=I,...,VI,1}^{\infty} \frac{[\psi_n'(0)]^2}{C_n} - 1 = 0, \qquad (2.156)$$

which yields the 1st identity of Eq. (2.154). Further, from the integral

$$I_2 = \frac{1}{2\pi i} \int_c \left\{ \frac{\alpha^2 \{\alpha T_1(i\alpha;2)\sinh \alpha h - K\cosh \alpha h\}}{\mathcal{G}(\alpha)\sinh \alpha h} - \frac{1}{\alpha} \right\} d\alpha, \qquad (2.157)$$

it can be derived that

$$I_2 = 2 \sum_{n=I,...,VI,1}^{\infty} \frac{k_n^2 [k_n T_1(ik_n;2)\sinh k_n h - K\cosh k_n h]}{\mathcal{G}'(k_n)\sinh k_n h} - 1$$

$$= \sum_{n=I,...,VI,1}^{\infty} \frac{[\psi_n'(h)]^2}{C_n} - 1 = 0, \qquad (2.158)$$

which proves the 2nd identity of Eq. (2.154). □

Corollary 7 *An equivalent form of the eigenfunctions $\psi_n(k_n, y)$ as in Eq. (2.148) is given by*

$$\psi_n(k_n,y) = \begin{cases} \left[\{T_2(ik_n;2) - s\}k_n \tanh k_n(H-h) - K \right] \dfrac{L_1(ik_n,y)}{L_1(ik_n,h)}, y \in (0,h), \\ -Ks\cosh k_n(H-y)/\cosh k_n(H-h), \qquad\qquad y \in (h,H). \end{cases}$$

This form of $\psi_n(k_n,y)$ is derived from Eq. (2.148) by using the alternate form of the dispersion relation as given by

$$\frac{(T_2(ik_n;2) - s)k_n - K\coth k_n(H-h)}{iL_1(ik_n,h)} = \frac{s}{\mathcal{D}(ik_n,h)}.$$

It may be noted that in the case of finite water depth for deriving the dispersion relation from the eigenfunctions as in Eq. (2.148), one has to use the interface boundary condition (2.137) while, the form of the eigenfunction in Corollary 2 can be used along with the boundary condition in Eq. (2.138) to derive the same dispersion relation.

2.6.4 Generalised expansion formulae

Here, the problem is also considered in the two-dimensional Cartesian coordinate system with the domain of consideration being the same as discussed in

Section 2.3. The spatial velocity potential $\phi(x, y)$ satisfies the Laplace equation (2.133) along with the boundary conditions in Eqs. (2.135)–(2.138). However, on the free surface at $y = 0$, it is assumed that the velocity potential ϕ satisfies a general type of boundary condition as given by

$$(\mathcal{L} + \mathcal{M})\phi = 0, \quad y = 0, \ 0 \leq x < \infty, \tag{2.159}$$

where $\mathcal{L}$, $\mathcal{M}$ are linear differential operators of the form

$$\mathcal{L} \equiv \sum_{j=0}^{l_0} c_j \frac{\partial^{2j}}{\partial x^{2j}} \left(\frac{\partial}{\partial y} \right), \quad \mathcal{M} \equiv \sum_{j=0}^{m_0} d_j \frac{\partial^{2j}}{\partial x^{2j}}, \tag{2.160}$$

c_j and d_j are known constants and l_0 and m_0 are positive integers with $m_0 \leq l_0$ (as discussed in Section 2.3). Further, the velocity potential will satisfy a Sommerfeld-type radiation condition as in Eq. (1.24). Further, in the case of finite water depth, the associated dispersion relation in k is given by

$$\mathcal{G}(k; l_0, m_0) \equiv \frac{K}{P(k; l_0)}$$

$$-\frac{Q(k; m_0)\{s \coth kh + \coth k(H - h)\} - k\{T_2(ik; 2) - s\}}{Q(k; m_0)\{s + \coth kh \coth k(H - h)\} - \{T_2(ik; 2) - s\}k \coth kh} = 0, \tag{2.161}$$

where $P(k; l_0) = \sum_{j=0}^{l_0} c_j(-1)^j k^{2j+1}$ and $Q(k; m_0) = \sum_{j=0}^{m_0} d_j(-1)^j k^{2j}$ are the characteristic polynomials associated with the differential operators $\mathcal{L}$ and $\mathcal{M}$, respectively.

The general form of velocity potential $\phi(x, y)$ satisfying Eq. (2.133) along with the boundary conditions in Eqs. (2.135), (2.137), (2.138) and (2.159) in the case of water of infinite depth is given by

$$\phi(x, y) = \sum_{n=I}^{2l_0+2} A_n F_n(k_n, y)e^{ik_n x} + \int_0^\infty A(\xi)L(\xi, y; l_0, m_0)e^{-\xi x}d\xi, \quad x > 0, \tag{2.162}$$

where $F_n(y)$ and $L(\xi, y)$ are given by

$$F_n(k_n, y) = \begin{cases} -\dfrac{iL_1(ik_n, y; l_0, m_0)}{K\mathcal{D}(ik_n, h; l_0, m_0)}, & 0 < y < h, \\ e^{-k_n(y-h)}, & h < y < H, \end{cases}$$

$$L(\xi, y; l_0, m_0) = \begin{cases} L_1(\xi, y; l_0, m_0) & \text{for } 0 < y < h, \\ L_2(\xi, y; l_0, m_0) & \text{for } h < y < \infty, \end{cases}$$

with

$$
\begin{aligned}
L_1(\xi, y; l_0, m_0) &= -iK\{P(i\xi; l_0)\cos\xi y - Q(i\xi; m_0)\sin\xi y\}, \\
L_2(\xi, y; l_0, m_0) &= L_1(\xi, y; l_0, m_0) - W(\xi, h; l_0, m_0)\cos\xi(y - h), \\
W(\xi, h; l_0, m_0) &= (1 - s)L_1(\xi, h; l_0, m_0) - \xi\{T_2(\xi; 2) - s\}\mathcal{D}(\xi, h; l_0, m_0), \\
\mathcal{D}(\xi, h; l_0, m_0) &= -iP(i\xi; l_0)\sin\xi h + Q(i\xi; m_0)\cos\xi h.
\end{aligned}
$$

Further, the eigenfunctions k_n satisfy the dispersion relation in k given by

$$
\mathcal{G}(k; l_0, m_0) \equiv \frac{P(k; l_0)[Q(k; m_0)(s\coth kh + 1) - k\{T_2(ik; 2) - s\}]}{Q(k; m_0)\{s + \coth kh\} - \{T_2(ik; 2) - s\}k\coth kh} - K = 0.
$$

(2.163)

Keeping the realistic nature of physical problems in mind, it is assumed that the dispersion relation in Eq. (2.163) has two distinct real positive roots k_I and k_{II} and the complex roots are $k_{III}, k_{IV}, ..., k_{2l_0+2}$ of the forms $a \pm ib$ and $-c \pm id$. Of the $2l_0 + 2$ roots, roots leading to boundedness of the velocity potential will be considered in the expansion formulae in Eq. (2.162). It may be noted that $\mathcal{D}(ik_n, h) \neq 0$ for all $0 < h < \infty$. Further, $F_n(k_n, y)$ satisfies

$$
\langle F_m, F_n\rangle = s\langle F_m, F_n\rangle_1 + \langle F_m, F_n\rangle_2 = C_n\delta_{mn},
$$

(2.164)

where

$$
\begin{aligned}
\langle F_m, F_n\rangle_1 &= \int_0^h F_m(y)F_n(y)dy \\
&+ \sum_{j=1}^{l_0} \frac{(-1)^j c_j}{Q(k_n; m_0)} \sum_{k=1}^{j} F_m^{2k-1}(0)F_n^{2(j-k)+1)}(0) \\
&+ \sum_{j=1}^{m_0} \frac{(-1)^{j+1}d_j}{P(k_n; l_0)} \sum_{k=1}^{j} F_m^{2k-2}(0)F_n^{2j-2k}(0),
\end{aligned}
$$

$$
\begin{aligned}
\langle F_m, F_n\rangle_2 &= \lim_{\epsilon \to 0} \int_h^{\infty} e^{-\epsilon y} F_m(y)F_n(y)dy \\
&+ \frac{D_2}{K}\left\{F_m'(h)F_n'''(h) + F_m'''(h)F_n'(h)\right\} - \frac{N_2}{K}F_m'(h)F_n'(h).
\end{aligned}
$$

$$
C_n = -\frac{\sinh k_n h[\psi_n'(0)]^2 \mathcal{G}'(k_n; l_0, m_0)\mathcal{D}(ik_n, h; l_0, m_0)}{2K^2 k_n^2},
$$

and δ_{mn} is the Knoneckar delta function. Further, the eigenfunctions $F_n(y)$ and $L(\xi, y; l_0, m_0)$ satisfy the generalised orthogonal mode-coupling relation as given by

$$
\langle F_m, L(\xi, y; l_0, m_0)\rangle = 0 \quad \text{for} \quad m = I, II, ..., 2l_0, \ \xi > 0.
$$

(2.165)

Using the orthogonal mode-coupling relation in Eq. (2.164), the unknowns A_n and $A(\xi)$ in Eq. (2.162) are obtained as

$$A_n = \frac{\langle \phi(x,y), F_n(k_n,y) \rangle}{C_n}, \quad A(\xi) = \frac{2\langle \phi(x,y), L(\xi,y;l_0,m_0) \rangle}{\pi \Delta(\xi,h;l_0,m_0)}, \quad (2.166)$$

where

$$\begin{aligned} \Delta(\xi,h;l_0,m_0) &= W^2(\xi,h;l_0,m_0) - 2W(\xi,h;l_0,m_0)L_1(\xi,h;l_0,m_0) \\ &+ K^2\{Q^2(i\xi;m_0) - P^2(i\xi;l_0)\}. \end{aligned}$$

The general form of velocity potential $\phi(x,y)$, satisfying Eq. (2.133) along with the boundary conditions in Eqs. (2.134), (2.137), (2.138) and (2.154) in the case of finite depth, is derived as

$$\phi(x,y) = \sum_{n=0,I}^{2l_0} A_n \psi_n(k_n,y)e^{ik_n x} + \sum_{n=1}^{\infty} A_n \psi_n(k_n,y)e^{-k_n x} \text{ for } x > 0, \quad (2.167)$$

where the eigenfunctions ψ_n can be obtained as

$$\psi_n(k_n,y) = \begin{cases} -\dfrac{i\tanh k_n(H-h)L_1(ik_n,y;l_0,m_0)}{K\mathcal{D}(ik_n,h;l_0,m_0)}, & 0 < y < h, \\ \dfrac{\cosh k_n(H-y)}{\cosh k_n(H-h)}, & h < y < H, \end{cases} \quad (2.168)$$

with the eigenvalues k_n satisfying the dispersion relation as given in Eq. (2.61). Keeping in mind that such boundary value problems are of physical interest and that our objective is to derive the expansion formulae which can be used for a wide class of problems in fluid structure interaction, it is assumed that the dispersion relation (2.161) has two distinct positive real roots k_n, $n = I, II$, k_n, $n = III, ..., 2l_0+1$ of the form $-a \pm ib$ and $-c \pm id$ and an infinite number of purely imaginary roots k_n, $n = 1, 2, ...$ of the form $k_n = ik_n$. However, a brief discussion about the behavior of wave numbers associated with progressive waves is provided in Section 1.4 of Chapter 1.

The eigenfunctions $\psi_n(k_n,y)$ satisfy the orthogonal relation

$$\begin{aligned} \langle \psi_m, \psi_n \rangle &= s\langle \psi_m, \psi_n \rangle_3 + \langle \psi_m, \psi_n \rangle_4 \\ &= C_n \delta_{mn}, \quad \text{for} \quad m,n = 0, I, II, ..., 2l_0, 1, 2, ..., \quad (2.169) \end{aligned}$$

where

$$\begin{aligned} \langle \psi_m, \psi_n \rangle_3 &= \int_0^h \psi_m(y)\psi_n(y)dy \\ &+ \sum_{j=1}^{l_0} \frac{(-1)^j c_j}{Q(k_n;m_0)} \sum_{k=1}^{j} \psi_m^{2k-1}(0)\psi_n^{2j-(2k-1)}(0) \\ &+ \sum_{j=1}^{m_0} \frac{(-1)^{j+1} d_j}{P(k_n;l_0)} \sum_{k=1}^{j} \psi_m^{2k-2}(0)\psi_n^{2j-2k}(0), \end{aligned}$$

$$\langle \psi_m, \psi_n \rangle_4 = \int_h^H \psi_m(y)\psi_n(y)dy$$

$$+ \frac{D_2}{K}\left\{\psi_m'(h)\psi_n'''(h) + \psi_m'''(h)\psi_n'(h)\right\} - \frac{Q_2}{K}\psi_m'(h)\psi_n'(h),$$

δ_{mn} being the Kroneckar delta function

$$C_n = -\frac{\sinh k_n h [\psi_n'(0)]^2 \mathcal{G}'(k_n; l_0, m_0)\mathcal{D}(ik_n, h; l_0, m_0)}{2K^2 k_n^2},$$

with $\mathcal{G}(k_n; l_0, m_0)$ being the same as in Eq. (2.166). Using the orthogonal mode-coupling relation as in Eq. (2.169), A_n is obtained as

$$A_n = \frac{\langle \phi(x,y), \psi_n(k_n, y) \rangle}{C_n}, \text{ for } n = 0, I, ..., 2l_0, 1, 2, ... \tag{2.170}$$

The expansion formulae and associated results derived in this section are meant for problems in two dimensions. From the general expansion formulae, expansion formulae associated with a class of physical problems arising in ocean engineering as discussed in Subsection 1.5. can be obtained as the particular cases (as in [10]). Further, as discussed in Section 2.3 in the case of single layer fluid, various expansion formulae and associated results in three dimensions can be easily derived (see [104]). Some of the applications of the expansion formulae which are of interest in ocean engineering are left to the readers as exercises.

2.7   Examples and exercises

Exercise 2.1 Derive the expansion formulae discussed in Section 2.3 in the cylindrical polar coordinate assuming that the floating elastic plate is circular in nature in water of finite and infinite water depth.

Exercise 2.2 Discuss the scattering of surface gravity waves by floating elastic circular plates of radius a in water of finite and infinite water depths in a homogeneous fluid of density ρ.

Exercise 2.3 Discuss the scattering of surface gravity waves by a finite and semi-infinite floating elastic plate in water of finite and infinite water depth.

Exercise 2.4 Discuss the scattering of flexural gravity waves by a submerged vertical breakwater placed below a floating elastic plate.

Exercise 2.5 Discuss the scattering of flexural gravity waves by a crack in the floating ice sheet in finite and infinite water depth in a two-layer fluid having a ice-covered free surface and an interface.

Exercise 2.6 Discuss the solution procedure to study the effect of a submerged horizontal flexible membrane on wave scattering by a finite floating flexible elastic plate.

Exercise 2.7 Derive the expansion formulae associated with gravity waves in the presence of surface and interfacial tension in finite/infinite water depths.

Exercise 2.8 Discuss the scattering of surface gravity waves by a horizontal submerged flexible plate/membrane for attenuation of wave height in finite and infinite water depths.

Exercise 2.9 Derive the expansion formulae associated with the wave motion in a two-layer fluid of finite water depth having free surface and interface in three dimensions and thus prove that the associated vertical eigenfunctions are linearly dependent.

Exercise 2.10 Generalise the expansion formulae associated with gravity wave interaction with floating flexible structure in the presence of submerged flexible plates/membrane in three dimensions in the cases of finite and infinite water depth and thus derive the associated orthogonal mode-coupling relations.

3

Green's function technique

3.1 Introduction

The mathematical modeling of a physical problem often gives rise to ordinary or partial differential equations satisfying a set of initial conditions and/or boundary conditions. It is very essential for one to understand the physical system one is trying to investigate and apply the most suitable approach for analysing the physical system in an efficient manner. Understanding the advantages and limitations of a method is very essential before applying the method. The mathematical models for a large class of physical problems lead to boundary value problems or initial boundary value problems associated with either ordinary differential equations (ODEs) or partial differential equations (PDEs). Methods of separation of variables are applied directly when the differential equation is homogeneous satisfying homogeneous boundary conditions. To the contrary, there is a large class of physical problems which lead to differential equations with a forcing term satisfying a set of initial conditions and/or nonhomogeneous boundary conditions. Such problems are often handled with the help of Green's function which is used to describe the influence of both nonhomogeneous boundary conditions and forcing terms. Often, it is convenient to provide the integral representation of the boundary value problem (BVP), which incorporates most of the boundary conditions by the suitable application of Green's identity and Green's function which is referred as Green's function technique, named after George Green (1793−1841). In this chapter, emphasis is given to the application of Green's function to derive the expansion formulae for wave structure interaction problems. Further, the utility of the expansion formulae is illustrated by analysing wave structure interaction problems in specific cases. Before going into the details of the application of Green's function in wave structure interaction problems, some of the basic characteristics and underlying procedures associated with Green's function techniques are discussed in brief.

3.1.1 The delta function and related distributions

The delta function has been widely used for solving a large class of physical problems after being introduced by P. A. M. Dirac (1902–1984) in his work on quantum mechanics prior to being defined as a generalised function by L. Schwartz (see [30], [80]). Let $\phi(\tau)$ be an arbitrary function which is continuous and infinitely many times differentiable at $t = \tau$. Then, the impulse function $\delta(t)$, also called the Dirac delta function or more precisely the delta distribution, is defined by the property

$$\phi(0) = \int_{-\infty}^{\infty} \delta(t)\phi(t)dt, \tag{3.1}$$

or more generally,

$$\phi(\tau) = \int_{-\infty}^{\infty} \delta(t - \tau)\phi(t)dt. \tag{3.2}$$

In particular, if $\phi(t) = 1$ and $t = 0$, then this gives

$$\int_{-\infty}^{\infty} \delta(\tau)d\tau = 1. \tag{3.3}$$

The n-th derivative $\delta^n(t)$ of the delta function is defined by

$$\phi^n(0) = (-1)^n \int_{-\infty}^{\infty} \delta^n(t)\phi(t)dt. \tag{3.4}$$

It is appropriate in this context to define the Heaviside step function $H(x)$, which is given by

$$H(x - \xi) = \begin{cases} 1, & x > \xi, \\ 0, & x < \xi. \end{cases} \tag{3.5}$$

Further, it may be noted that

$$\int_{-\infty}^{\infty} H'(x - \xi)h(x)dx = - \int_{-\infty}^{\infty} H(x - \xi)h'(x)dx = \int_{\xi}^{\infty} h'(x)dx = h(\xi), \tag{3.6}$$

which gives

$$H'(x - \xi) = \delta(x - \xi). \tag{3.7}$$

In mathematical physics and engineering, often the input to a linear system is an impulse at time $t = t_0$ and is understood as $\delta(t - t_0)$.

3.1.2   Green's function technique for ODE

Green's function technique is illustrated here through a simple boundary value problem associated with a loaded string. The problem is mathematically represented as a BVP associated with the function $u(x)$ which satisfies

$$u''(x) = \phi(x); \quad u(0) = u(1) = 0, \tag{3.8}$$

where $\phi(x)$ is assumed to be known. It is often helpful to recast the mathematical problem in terms of physical terminology. Accordingly, $u(x)$ is regarded as the static deflection of a string, stretched under unit tension between two fixed points and subjected to a force distribution $\phi(x)$. In the present problem, Green's function $G(x, \xi)$ satisfies

$$G_{xx} = \delta(x - \xi) \qquad (3.9)$$

subject to the boundary conditions

$$G(0, \xi) = G(1, \xi) = 0. \qquad (3.10)$$

Suppose

$$G(x, \xi) = U(x, \xi) + g(x, \xi), \qquad (3.11)$$

where U satisfies

$$U_{xx} = \delta(x - \xi), \qquad (3.12)$$

subject to no particular boundary conditions which on integration yields

$$U = (x - \xi)H(x - \xi), \qquad (3.13)$$

where H is the Heaviside step function as defined in Eq. (3.7). Thus, U does contain the basic singularity, namely, the "kink" at $\xi = x$ but does not satisfy the boundary conditions (3.10) since $U(0, \xi) = 0, U(1, \xi) = 1 - \xi$. Thus, g satisfies the homogeneous equation

$$g_{xx} = 0 \qquad (3.14)$$

subject to the boundary conditions which are such that the combination $U + g$ satisfies Eq. (3.10). Thus,

$$G(0, \xi) = 0 \Rightarrow g(0, \xi) = 0. \qquad (3.15)$$

Further,

$$G(1, \xi) = 0 \Rightarrow g(1, \xi) = \xi - 1. \qquad (3.16)$$

Thus, the solution of the boundary value problem in $g(x, \xi)$ is given by

$$g(x, \xi) = (\xi - 1)x, \qquad (3.17)$$

which yields

$$G(x, \xi) = (x - \xi)H(x - \xi) + (\xi - 1)x, \qquad (3.18)$$

where $H(x)$ is the Heaviside function as defined in Eq. (3.5). In this case $G(x, \xi)$ is the deflection, as a function of x, not due to the load distribution ϕ but rather due to a point load of unit strength $\delta(x - \xi)$ acting at the point $x - \xi$. The above expression for Green's function is rewritten as

$$G(x, \xi) = \begin{cases} (\xi - 1)x, & x \leq \xi, \\ (x - 1)\xi, & x \geq \xi. \end{cases} \qquad (3.19)$$

It is clear that that G is symmetric, that is, $G(x, \xi) = G(\xi, x)$, which says that the deflection at x due to a unit load at ξ is equal to the deflection at ξ due to a unit load at x. Thus, the solution of the differential equation (3.8) is given by

$$u(x) = \int_0^1 G(x, \xi)\phi(\xi)d\xi = (x - 1) \int_0^x \xi\phi(\xi)d\xi + x \int_x^1 (\xi - 1)\phi(\xi)d\xi. \quad (3.20)$$

In the above solution for $u(x)$, $G(x, \xi)\phi(\xi)d\xi$ is called the deflection due to an incremental load $\phi(\xi)d\xi$ at x, so that $u(x)$ represents the superposition of the resulting incremental deflections. Green's function G is often called the influence function and is the kernel of the integral in (3.20). Although this is a simple example to illustrate the splitting technique and to solve a boundary value problem associated with an ordinary differential equation, the concept to solve complicated problems follows in a similar manner. The same approach can easily be extended to solve an ordinary differential equation of higher order and even can be converted to an integral equation (see [41] and [58] for further details).

In case the forcing function $\phi(x)$ is replaced by $\phi(x, u(x))$, which in itself is a differential operator, the integral (3.20) involves the unknown function $u(x)$ along with its derivatives and in such a situation, (3.20) is an integral or integro-differential equation in $u(x)$. In the derived integral equation, $G(x, \xi)$ is called the kernel of the integral equation. Through this process, often used to solve complex equations, simple Green's functions, which can be easily computed, are used to convert the original differential equation into an integral equation.

In general, it is not always possible to obtain a closed-form solution for an integral equation. In such a situation, one has to solve the integral equation numerically. As noted earlier, in the numerical computation, one discretises the boundary of the region to derive the full solution details will not be considered here.

3.1.3   Green's function method for PDE

A large class of physical problems of mathematical physics and engineering related to problems of fluid mechanics, solid mechanics, fluid structure interactions, electromagnetic theory, acoustics, etc., leads to BVPs associated with Laplace or Helmholtz equations in the finite or infinite domain. Often it is more appropriate to convert the BVP into an integral or integro-differential equation whose approximate solution can be obtained in an efficient manner. For example, most of the problems associated with wave structure interaction in marine environments are tackled by considering the respective boundary integral equation formulation. In this subsection, solutions of boundary value problems associated with reduced wave equations satisfying higher order boundary conditions are obtained directly by the suitable application of Green's integral theorem and the associated Green's function which is the same as the expansion formulae derived in the previous subsection. One of the primary objectives of the technique is to find Green's function, which satisfies a much simpler BVP than the original problem at hand, the details of which are described next.

Green's function is an integral kernel that can be used to solve a nonhomogeneous differential equation with boundary conditions and a key component in obtaining an integral representation of the solution associated with a BVP. Green's function $G(x, y; \xi, \eta)$ is rewritten in two parts as

$$G(x, y; \xi, \eta) = U(x, y; \xi, \eta) + g(x, y; \xi, \eta), \tag{3.21}$$

where U is a particular solution of the associated BVP, which need not satisfy the required boundary conditions, and g is a solution of the homogeneous differential equation, such that the combination $U + g$ does satisfy those boundary conditions. Here, U is referred as the principal solution which is also known as the fundamental solution, an elementary solution, often called a free-space Green's function in the literature. It contains the basic singularity of Green's function. This process of splitting G into two parts is often referred to as the splitting technique. In the context of the present work, the fundamental solutions associated with the famous elliptic operators are highlighted without details which can be found in various textbooks available in the literature (see [31], [148]).

In two dimensions, the fundamental solution associated with the two-dimensional Laplace operator $U(x, y; \xi, \eta)$ satisfies

$$\nabla^2 U = \delta(x - \xi, y - \eta) = \delta(x - \xi)\delta(y - \eta), \tag{3.22}$$

which may be physically regarded as the potential induced at a field point (ξ, η) by a point mass of unit strength at (x, y) in a gravitational field and $\nabla^2 \equiv \partial^2/\partial x^2 + \partial^2/\partial y^2$. The solution of Eq. (3.22) in the two-dimensional plane is given by

$$U(x, y; \xi, \eta) = \frac{1}{2\pi} \ln r, \tag{3.23}$$

where $r = \sqrt{(x - \xi)^2 + (y - \eta)^2}$. On the other hand, in three dimensions, the fundamental solution $U(x, y, z; \xi, \eta, \zeta)$ satisfies

$$\nabla^2 U = \delta(x - \xi, y - \eta, z - \zeta) = \delta(x - \xi)\delta(y - \eta)\delta(z - \zeta), \tag{3.24}$$

with the solution

$$U = (-1)/(4\pi r), \tag{3.25}$$

where $r = \sqrt{(x - \xi)^2 + (y - \eta)^2 + (z - \zeta)^2}$. Further, the fundamental solutions associated with the two-dimensional Helmholtz equation which satisfies

$$\nabla^2 U + k^2 U = \delta(x - \xi, y - \eta) = \delta(x - \xi)\delta(y - \eta) \tag{3.26}$$

is given by

$$u(x, y; \xi, \eta) = -\frac{i}{4} H_0^{(1)}(kr),$$

while the fundamental solution associated with the three-dimensional Helmholtz equation which satisfies

$$\nabla^2 U + k^2 U = \delta(x - \xi, y - \eta, z - \zeta) = \delta(x - \xi)\delta(y - \eta)\delta(z - \zeta), \qquad (3.27)$$

where ∇^2 is the three-dimensional Laplacian operator, is given by

$$u(x, y, z; \xi, \eta, \zeta) = \frac{e^{ikr}}{4\pi r} \qquad (3.28)$$

with r being the same as in Eq. (3.25). Finally, the fundamental solution $u(x, y; \xi, \eta)$ associated with the reduced wave equation which satisfies

$$\nabla^2 U - k^2 U = \delta(x - \xi, y - \eta) = \delta(x - \xi)\delta(y - \eta), \qquad (3.29)$$

is given by

$$u(x, y) = K_0(kr), \qquad (3.30)$$

where r is the same as in Eq. (3.23).

It may be noted that in the case of the Laplace differential operator, the two independent variables ξ and η appear symmetrically in the PDEs. This implies that U could be considered to be symmetric about the singular point (x, y). Green's function technique associated with the Laplace equation is based on the famous Green's identities. The integral identities

$$\int_R u\nabla^2 v \, dV = -\int_R (\nabla u).(\nabla v)dV + \int_S u\frac{\partial v}{\partial n} dS \qquad (3.31)$$

and

$$\int_R (u\nabla^2 v - v\nabla^2 u)dV = \int_S \left\{ u\frac{\partial v}{\partial n} - v\frac{\partial u}{\partial n} \right\} dS, \qquad (3.32)$$

where the functions u and v have partial derivatives of the second order which are continuous in the bounded region R and S is the boundary of R, while n denotes the outward normal to S, are known as the Green's first and second identities, respectively. The surface S is smooth (regular) as in [31] and the operator ∇^2 is the Laplacian operator in three dimensions. In particular, if ϕ and ψ both satisfy the Laplace equation ($\nabla^2\phi = \nabla^2\psi = 0$), then Green's second identity (setting $u = \phi$ and $v = \psi$) yields

$$\int\int_S \left\{ \phi\frac{\partial \psi}{\partial n} - \psi\frac{\partial \phi}{\partial n} \right\} dS = \int\int\int_R \left\{ \phi\nabla^2\psi - \psi\nabla^2\phi \right\} dV = 0. \qquad (3.33)$$

It can be easily derived that if G is Green's function for the Laplacian operator and ϕ is the velocity potential which satisfies the Laplace equation, then the following two identities will hold. In three dimensions, if a closed domain of surface S is considered with $\hat{n}$ being the outward normal vector, then

$$\int\int_S \left\{ \phi\frac{\partial G}{\partial n} - G\frac{\partial \phi}{\partial n} \right\} dS = \begin{cases} 0, & (x, y, z) \text{ outside S,} \\ 2\pi\phi(x, y, z), & (x, y, z) \text{ on S,} \\ 4\pi\phi(x, y, z), & (x, y, z) \text{ inside S.} \end{cases} \qquad (3.34)$$

On the other hand, in the case of two dimensions,

$$\int_S \left\{ \phi \frac{\partial G}{\partial n} - G \frac{\partial \phi}{\partial n} \right\} dS = \begin{cases} 0, & (x,y) \text{ outside S}, \\ \pi \phi(x,y), & (x,y) \text{ on S}, \\ 2\pi \phi(x,y), & (x,y) \text{ inside S}. \end{cases} \tag{3.35}$$

Here, Green's function technique to solve the two-dimensional Laplace equation in bounded and unbounded domains is discussed in detail. The solution of the Laplace equation $\nabla^2 u = 0$ in R will be determined by using the appropriate boundary conditions either $u = f$ on S or $\partial u / \partial n = g$ on S, where f and g are two given continuous functions on S.

In the case of a two-dimensional Dirichlet problem, using Green's second identity as in Eq. (3.32), it can be easily derived that

$$u(x,y) = -\int_S f(s) \frac{\partial G_1}{\partial n} ds, \tag{3.36}$$

where G_1 satisfies

$$\nabla^2 G_1 = \delta(x - \xi)\delta(y - \eta), \tag{3.37}$$

along with the boundary condition $G_1 = 0$ on S and in the case of a Neumann problem

$$u(x,y) = -\int_S g(s) G_2 ds, \tag{3.38}$$

where G_2 satisfies the Eq. (3.37) along with the boundary condition $\partial G_2 / \partial n = 0$ on S. If the region is the half plane or the quarter plane, for the calculation of G, the problem can be converted to a differential equation by using the appropriate Fourier transform. Then the differential equation can be solved easily.

3.2 Green's function technique for expansion formulae

A large class of fluid structure interaction problems gives rise to boundary value problems having higher order boundary conditions in one or more of its boundaries in a half plane or infinite strip. Using the symmetric behavior of the domain of considerations, often these problems are reduced to problems in a quarter plane or in a semi-infinite strip satisfying a Dirichlet- or Neumann-type nonhomogeneous boundary condition in one of the boundaries apart from satisfying higher order boundary conditions in the other boundary. In addition, there are physical problems which give rise to boundary value problems satisfying a Dirichlet- or Neumann-type nonhomogeneous boundary condition on one of the boundaries.

In this section, considering the boundary value problems having a boundary condition of the Dirichlet type, the Fourier sine transform is utilized to find the expansion formulae in the case of water of infinite depth. However, for the boundary value problem having a Neumann-type boundary condition, the Fourier-cosine transform can be used. The uniqueness of the solutions depends on the far field behavior in addition to the edge behaviors. The edge behaviors are obtained from the physical requirement of the boundary value problem, whose details are discussed in the physical problems.

3.2.1   General boundary value problem

For the sake of clarity, the general boundary value problems associated with physical problems in the broad area of fluid structure interaction problems are discussed in a quarter plane and in a semi-infinite strip. The boundary value problems will be discussed in the three-dimensional Cartesian coordinate system with the motion being symmetric along the z-axis such that the total velocity potential $\Phi(x,y,z,t) = Re\{\phi(x,y)e^{i\nu_0 z - i\omega t}\}$, which is very common while dealing with wave structure interaction problems associated with oblique waves. This assumption reduces the three-dimensional problem to two-dimensional in the case of a half/quarter plane (referred as infinite water depth) and an infinite/semi-infinite strip (referred to as finite water depth). Assume that the fluid occupies the region $0 < x < \infty, 0 < y < h$ in the case of finite water depth and the region $0 < x < \infty, 0 < y < \infty$ in the case of infinite water depth. With similar assumptions on the nature of fluid and its motion as discussed in Chapter 2, in this case also, the spatial velocity potential $\phi(x,y)$ satisfies the governing PDE

$$(\nabla^2 - \nu_0^2)\phi = 0 \quad \text{in the fluid domain,} \tag{3.39}$$

where $\nabla^2 = \partial^2/\partial x^2 + \partial^2/\partial y^2$ along with structural boundary conditions on the mean free surface as given by

$$D\left(\frac{\partial^2}{\partial x^2} - \nu_0^2\right)^2 \frac{\partial \phi}{\partial y} + Q\left(\frac{\partial^2}{\partial x^2} - \nu_0^2\right)\frac{\partial \phi}{\partial y} + \frac{\partial \phi}{\partial y} + K\phi = 0, \quad y = 0, x > 0, \tag{3.40}$$

with D, Q and K being the same as defined in Eqs. (1.95) and (2.24). The bottom boundary condition is given by

$$\phi_y = 0 \quad \text{on } y = h \text{ (finite depth),} \tag{3.41}$$

$$\phi, |\nabla \phi| \rightarrow 0 \quad \text{as } y \rightarrow \infty \text{ (infinite depth).} \tag{3.42}$$

Further, on the vertical boundary at $x = 0$, it is assumed that the velocity potential $\phi(x,y)$ satisfies

$$\phi(x,y) = u(y) \quad \text{on} \quad x = 0. \tag{3.43}$$

In addition, the velocity potential satisfies the radiation condition given by

$$\phi(x,y) \rightarrow A_0 f_0(y)e^{i\mu_0 x}, \quad \text{as } x \rightarrow \infty, \tag{3.44}$$

where $\qquad f_0(y) \rightarrow \begin{cases} \dfrac{\cosh k_0 (h-y)}{\cosh k_0 h}, & \text{in the case of finite water depth,} \\ e^{-k_0 y}, & \text{in the case of infinite water depth} \end{cases}$

and $\mu_0^2 + \nu_0^2 = k_0^2$ with k_0 satisfying the dispersion relation in k given by

$$F_{pd}(k) \equiv \begin{cases} (Dk^4 - Qk^2 + 1)k \tanh kh = K, & \text{in finite depth,} \\ (Dk^4 - Qk^2 + 1)k = K, & \text{in infinite depth.} \end{cases} \tag{3.45}$$

The uniqueness of the aforementioned boundary value problem will require certain edge conditions to be prescribed at the edges and depends on the type of physical problem and are deferred here. While illustrating various physical problems in a later section, appropriate edge conditions will be described. In Chapter 1 and Chapter 2, various types of edge conditions are discussed in brief and are not discussed here to avoid repetition. It may be noted that expansion formulae associated with the Laplace equation in two and three dimensions are derived in Chapter 2 based on the application of eigenfunction expansion method in the case of finite water depth and Fourier sine transform in the case of infinite water depth. In this chapter, the governing equation will be the two-dimensional reduced wave equation as in Eq. (3.39). In this section, the expansion formulae will be derived based on the application of Green's function technique. The Fourier sine transform along the horizontal axis is used to reduce the two-dimensional reduced wave equation to Sturm–Liouville-type boundary value problems in the cases of both finite and infinite water depth, which are solved by Green's function technique. Further, the regularity criteria of the Fourier sine transform is used to derive the wave amplitude at the far field.

3.2.2 Expansion formulae in the case of infinite water depth

In this subsection, the expansion formula for the velocity potential satisfying the governing equation (3.39) and the boundary conditions in Eq. (3.40) and (3.42)–(3.45) is obtained in the case of water of infinite depth. In order to derive the expansion formula, writing

$$\phi(x,y) = A_0 f_0(y) e^{i\mu_0 x} + \psi(x,y) \tag{3.46}$$

and taking the Fourier sine transform of $\psi(x,y)$ as defined by

$$\hat{\psi}_s(\xi, y) = \int_0^\infty \psi(x,y) \sin \xi x \, dx, \tag{3.47}$$

the boundary value problem in $\psi(x,y)$ is converted to a boundary value problem of the Sturm–Liouville-type associated with an ordinary differential equation in $\hat{\psi}_s(\xi, y)$. The boundary value problem in $\hat{\psi}_s(\xi, y)$ is given by

$$\frac{d^2 \hat{\psi}_s}{dy^2} - (\xi^2 + \nu_0^2) \hat{\psi}_s = v(y), \quad y > 0, \tag{3.48}$$

subject to the boundary conditions

$$\{D(\xi^2 + \nu_0^2)^2 - Q(\xi^2 + \nu_0^2) + 1\}\frac{d\hat{\psi}_s}{dy} + K\hat{\psi}_s = \hat{f}(\xi), \quad \text{on} \quad y = 0, \quad (3.49)$$

$$\frac{d\hat{\psi}_s}{dy} = 0, \quad \text{as} \quad y \to \infty, \quad (3.50)$$

where
$$v(y) = -\xi\{u(y) - A_0 f_0(y)\}$$
$$\hat{f}(\xi) = \xi\Big[\{D(\xi^2 + \nu_0^2) - Q\}\phi_y + D - \phi_{yyy}\Big]_{(x,y)=(0,0)}$$
$$+ A_0\xi k_0\{D(\xi^2 + \nu_0^2 + k_0^2) - Q\}.$$

The solution of Eq. (3.48) satisfying the boundary conditions in Eqs. (3.49) and (3.50) is derived by Green's function technique and is given by

$$\hat{\psi}_s(\xi, y) = \frac{-S(\xi, y)}{\sqrt{\xi^2 + \nu_0^2}\,H(\xi)}\int_0^\infty e^{-\sqrt{\xi^2 + \nu_0^2}\,t}g(\xi, t)dt$$

$$- \frac{1}{\sqrt{\xi^2 + \nu_0^2}}\int_0^y \sinh\{\sqrt{\xi^2 + \nu_0^2}(t - y)\}g(\xi, t)dt + \frac{\hat{f}(\xi)}{K}, \quad (3.51)$$

where

$$g(\xi, y) = \xi\{A_0 f_0(y) - u(y)\} + \{\xi^2 + \nu_0^2\}\hat{f}(\xi)/K,$$

$$H(\xi) = \sqrt{\xi^2 + \nu_0^2}\{1 - Q(\xi^2 + \nu_0^2) + D(\xi^2 + \nu_0^2)^2\} - K,$$

$$S(\xi, y) = \sqrt{\xi^2 + \nu_0^2}\{1 - Q(\xi^2 + \nu_0^2) + D(\xi^2 + \nu_0^2)^2\}\cosh\sqrt{\xi^2 + \nu_0^2}\,y$$

$$- K\sinh\sqrt{\xi^2 + \nu_0^2}\,y.$$

Assume that $H(\xi)$ has one positive real root at $\xi = \mu_0$ (say) and four complex roots at $\xi = \mu_n(n = I, ..., IV)$ of the form $\pm\alpha \pm i\beta$ in the context of the present analysis (as in Section 2.3). The positive real root of $H(\xi) = 0$ at $\xi = \mu_0$ ensures that the function $\hat{\psi}_s(\xi, y)$ has a singularity at $\xi = \mu_0$ on the positive real axis. However, $\hat{\psi}_s(\xi, y)$ being the Fourier sine transform of a function cannot have a singularity on the line $\xi > 0$, which ensures (as in [59])

$$\lim_{\xi \to \mu_0}(\xi - \mu_0)\hat{\psi}_s(\xi, y) = 0. \quad (3.52)$$

On simplifying, Eq. (3.52) gives rise to the constant A_0 as given by

$$A_0 = \frac{1}{C_0}\int_0^\infty u(t)e^{-k_0 t}dt + \frac{k_0\big[D\phi_{yyy} + (Q - Dk_0^2)\phi_y\big]}{C_0 K}\bigg|_{(x,y)=(0,0)}, \quad (3.53)$$

where $C_0 = F'_{pd}(k_0)/2K$ and is similar to the one defined in Eq. (2.31). Next, the Fourier sine inversion of $\hat{\psi}_s(\xi, y)$ as in Eq. (3.51) yields

$$\psi(x, y) = \frac{2}{\pi} \int_0^\infty \hat{\psi}_s(\xi, y) \sin \xi x d\xi. \tag{3.54}$$

Now, by writing $\sin \xi x$ as $(e^{i\xi x} - e^{-i\xi x})/2i$ in (3.54) and rotating the contour along the positive imaginary axis for the integral involving $e^{i\xi x}$ and along the negative imaginary axis for the integral involving $e^{-i\xi x}$ and using the Cauchy residue theorem of complex function theory, $\psi(x, y)$ in Eq. (3.54) is obtained as

$$\begin{aligned}
\psi(x, y) &= A_I f_I(y) e^{i\sqrt{k_I^2 - v_0^2}\,x} + A_{II} f_{II}(y) e^{-i\sqrt{k_{II}^2 - v_0^2}\,x} \\
&+ \frac{2}{\pi} \int_0^\infty \frac{A(\xi) M(\xi, y) e^{-\sqrt{\xi^2 + v_0^2}\,x} d\xi}{\Delta(\xi)},
\end{aligned} \tag{3.55}$$

where A_n for $n = I, II, M(\xi, y), A(\xi)$ and $\Delta(\xi)$ are given by

$$\begin{aligned}
A_n &= \frac{1}{C_n} \int_0^\infty u(t) f_n(t) dt \\
&+ \frac{k_n}{C_n K} \Big\{ D\phi_{yyy} + (Q - Dk_n^2)\phi_y \Big\}\Big|_{(x,y)=(0,0)}, \\
A(\xi) &= \int_0^\infty u(t) M(\xi, t) dt + \xi \Big\{ (Q + D\xi^2)\phi_y - D\phi_{yyy} \Big\}\Big|_{(x,y)=(0,0)}, \\
f_n(y) &= e^{-k_n y}, \ M(\xi, y) = \xi(1 - Q\xi^2 + D\xi^4) \cos \xi y - K \sin \xi y, \\
C_n &= F'_{pd}(k_n)/2K, \ \Delta(\xi) = \xi^2(1 + Q\xi^2 + D\xi^4)^2 + K^2,
\end{aligned}$$

and k_I and k_{II} are the complex roots of equation (3.45) with positive real parts of the forms $\alpha \pm i\beta$ and $\mu_n = \sqrt{k_n^2 - v_0^2}$ for $n = 0, I, II$. Substituting for $\psi(x, y)$ in the relation (3.47), the required expansion formula as desired is obtained. It may be noted that since k_{III} and k_{IV} are the roots of the equation (3.45) having negative real parts does not contribute to the expansion formula which is also clear from the boundedness criteria of the velocity potential.

It may be noted that for $v_0 = 0$, the expansion formulae reduce to the expansion formula associated with the two-dimensional Laplace equation and are similar to the one derived in Chapter 2. Further, it may be noted that the expansion formula depends on the behavior of ϕ, ϕ_y and ϕ_{yyy} at origin, which will depend on the edge behavior of the physical problem under consideration.

3.2.3   Expansion formulae in the case of finite water depth

In order to find $\phi(x, y)$ satisfying the governing equation (3.39) and the boundary conditions (3.40), (3.41), (3.43) and (3.44) in the case of finite water depth, set

$$\phi(x, y) = A_0 f_0(y) e^{i\mu_0 x} + \psi(x, y). \tag{3.56}$$

Using the Fourier sine transform of $\psi(x, y)$ as defined in Eq. (3.47), the boundary value problem is converted into a boundary value problem of the Sturm–Liouville-type associated with an ordinary differential equation (ODE) in $\hat{\psi}_s(\xi, y)$. In this case, $\hat{\psi}_s(\xi, y)$ satisfies the Eq. (3.48) subject to the boundary condition (3.49) with

$$
\begin{aligned}
v(y) &= -\xi\{u(y) - A_0 f_0(y)\}, \\
\hat{f}(\xi) &= \xi\{(D(\xi^2 + \nu_0^2) - Q)\phi_y + D\phi_{yyy}\}_{(x,y)=(0,0)} \\
&\quad + A_0 \xi k_0 \{D(\xi^2 + k_0^2 + \nu_0^2) - Q\} \tanh kh.
\end{aligned}
$$

Further, $\hat{\psi}_s(\xi, y)$ satisfies the boundary condition

$$
\frac{d\hat{\psi}_s}{dy} = 0, \quad \text{on} \quad y = h. \tag{3.57}
$$

The solution of the boundary value problem in $\hat{\psi}_s(\xi, y)$ is given by

$$
\begin{aligned}
\hat{\psi}_s(\xi, y) &= \frac{-S(\xi, y)}{\sqrt{\xi^2 + \nu_0^2}\, H(\xi)} \int_0^h \cosh\sqrt{\xi^2 + \nu_0^2}(h - t) g(\xi, t) dt \\
&\quad - \frac{1}{\sqrt{\xi^2 + \nu_0^2}} \int_0^y \sinh\sqrt{\xi^2 + \nu_0^2}(t - y) g(\xi, t) dt + \frac{\hat{f}(\xi)}{K},
\end{aligned}
$$

$$
\begin{aligned}
\text{where } S(\xi, y) &= \sqrt{\xi^2 + \nu_0^2}\{1 - Q(\xi^2 + \nu_0^2) + D(\xi^2 + \nu_0^2)^2\} \cosh\sqrt{\xi^2 + \nu_0^2}\, y \\
&\quad - K \sinh\sqrt{\xi^2 + \nu_0^2}\, y, \\
H(\xi) &= \sqrt{\xi^2 + \nu_0^2}\{1 - Q(\xi^2 + \nu_0^2) + D(\xi^2 + \nu_0^2)^2\} \sinh\sqrt{\xi^2 + \nu_0^2}\, h \\
&\quad - K \cosh\sqrt{\xi^2 + \nu_0^2}\, h, \\
g(\xi, y) &= -\xi\{u(y) - A_0 f_0(y)\} + (\xi^2 + \nu_0^2)\hat{f}(\xi)/K.
\end{aligned}
$$

It is observed that $H(\xi) = 0$ has a real positive root at $\xi = \mu_0$, which suggests that the function $\hat{\psi}_s(\xi, y)$ has a singularity at $\xi = \mu_0$ on the positive real axis. However, $\hat{\psi}_s(\xi, y)$ being the Fourier sine transform of a function cannot have a singularity on the line $\xi > 0$, which yields

$$
\lim_{\xi \to \mu_0} (\xi - \mu_0)\hat{\psi}_s(\xi, y) = 0. \tag{3.58}
$$

On simplifying, Eq. (3.58) yields

$$
\begin{aligned}
A_n &= \frac{1}{C_n} \int_0^h u(t) f_n(t) dt + \frac{k_n \tanh k_n h}{K C_n}\{ - D\phi_{yyy} \\
&\quad + (Q - Dk_n^2)\phi_y\}|_{(x,y)=(0,0)}, \text{ for } n = 0, \tag{3.59}
\end{aligned}
$$

with C_n being the same as defined in Eq. (3.55), and $f_n(t)$ is obtained from $f_0(t)$ by putting $k_0 = k_n$ for $n = I, II$ and $k_0 = ik_n$ for $n = 1, 2, ...$ in Eq. (3.44) in the case of finite water depth. The Fourier sine inversion of $\hat{\psi}_s(\xi, y)$ yields

$$\psi(x, y) = \frac{2}{\pi} \int_0^\infty \hat{\psi}_s(\xi, y) \sin \xi x d\xi. \tag{3.60}$$

Now, by writing $\sin \xi x$ as $(e^{i\xi x} - e^{-i\xi x})/2i$ in (3.60) and rotating the contour along the positive imaginary axis for the integral involving $e^{i\xi x}$ and along the negative imaginary axis for the integral involving $e^{-i\xi x}$ and using the Cauchy residue theorem of complex function theory, (3.60) can be rewritten as

$$\psi(x, y) = A_I f_I(y) e^{i\mu_I x} + A_{II} f_{II}(y) e^{-i\mu_{II} x} + \sum_{n=1}^\infty A_n f_n(y) e^{-\sqrt{k_n + v_0^2} x}, \tag{3.61}$$

where A_n is as in (3.59) with $k_n = k_n$ for $n = I, II$ and $k_n = ik_n$ for $n = 1, 2, ...$ Substituting for $\psi(x, y)$ from relation (3.61) and A_n from relation (3.59) in the relation (3.56), the required expansion formula as desired is obtained.

NB: It may be noted that in Chapter 2, the Fourier sine transform in the vertical boundary is used (along the y-axis) in the case of water of infinite depth to derive the expansion formulae. On the other hand, the eigenfunction expansion method was used to obtain the expansion formulae in the case of finite water depth. To obtain the unknown coefficients, the orthogonal mode-coupling relation as appropriate is used assuming that the eigenfunctions are complete. However, in the present study, the general solution procedure is described in the case of water of infinite depth with suitable application of the Cauchy residue theorem. The advantage of the present approach is that it is independent of fluid depth and it does not require the utilization of the orthogonal mode-coupling relation for the determination of the unknown coefficients in the expansion formulae. The derivation of the expansion formulae is simple and straightforward.

3.3 Green's theorem for expansion formulae

In this section, expansion formulae for wave structure interaction problems associated with reduced wave equations will be derived using Green's identities and the appropriate choice of Green's functions. Construction of Green's function for a class of problems in itself brings lots of challenges to the scientific community. There are various forms of Green's function that exist for free surface wave problems. Some of the well-known results associated with the free surface Green's function can be found in [82], [129] and [142]. In the present context, a simple procedure to derive Green's function is discussed and then the expansion formulae are derived with the help of the derived Green's function based on Green's identity.

3.3.1   Derivation of Green's functions

Green's function $G(x, y; \xi, \eta)$ satisfies the reduced wave equation

$$\left(\frac{\partial^2}{\partial x^2} + \frac{\partial^2}{\partial y^2} - \nu_0^2\right)G = 0 \tag{3.62}$$

in the fluid domain except at (ξ, η) along with the free surface boundary condition as given by the linearised ice/plate covered boundary condition on the mean free surface in the presence of compressive force as in Eq. (1.114) (neglecting $\rho_p d$) and is given by

$$D\frac{\partial^5 G}{\partial y^5} - Q\frac{\partial^3 G}{\partial y^3} + \frac{\partial G}{\partial y} + KG = 0, \quad \text{on} \quad y = 0, \quad 0 < x < \infty, \tag{3.63}$$

which is the equivalent form of the boundary condition in Eq. (3.40). The velocity potential $\phi(x, y)$ satisfies the bottom boundary condition (3.41) or (3.42) as appropriate for the fluid of finite or infinite depths. The radiation condition in this case is of the form

$$G \rightarrow \begin{cases} Ae^{-k_0 y + i\mu_0 |x-\xi|}, & \text{as} \quad |x-\xi| \rightarrow \infty, \quad \text{(infinite depth)}, \\ Bf_0(y)e^{i\mu_0 |x-\xi|}, & \text{as} \quad |x-\xi| \rightarrow \infty, \quad \text{(finite depth)}, \end{cases} \tag{3.64}$$

where $f_0(y) = \cosh k_0(h - y)/\cosh k_0 h$, A and B are constants to be determined and k_0 satisfies the respective dispersion relations in (2.26) in the case of infinite and finite water depths with $k_0^2 = \nu_0^2 + \mu_0^2$. In addition, $G(x, y; \xi, \eta)$ satisfies the condition

$$G(x, y; \xi, \eta) \sim K_0(\nu_0 r), \quad \text{as} \quad r = \sqrt{(x-\xi)^2 + (y-\eta)^2} \rightarrow 0, \tag{3.65}$$

where $K_0(\nu_0 r)$ is the zeroth order modified Bessel function of the second kind. From the integral form of $K_0(\nu_0 r)$ given by (as in [88])

$$K_0(\nu_0 r) = \int_0^\infty \frac{e^{-\sqrt{\nu_0^2 + \zeta^2}|x-\xi|} \cos \zeta(y-\eta) d\zeta}{\sqrt{\nu_0^2 + \zeta^2}}, \tag{3.66}$$

with r being the same as in Eq. (3.23), it can be easily derived that

$$\lim_{x \to 0} \frac{\partial G(x, y; 0, \eta)}{\partial x} = \frac{1}{2}\delta(y - \eta), \tag{3.67}$$

where $\delta(y)$ is the delta function as defined in Eq. (3.1). Utilising the expansion formulae as discussed in the previous section, Green's function $G(x, y; \xi, \eta)$ in the case of infinite depth can be expressed as

$$G(x,y;\xi,\eta) = \sum_{n=0,I}^{II} A_n e^{i\mu_n|x-\xi|-k_n y}$$
$$+ \frac{2}{\pi} \int_0^\infty \frac{A(\xi)M(\xi,y)e^{-\sqrt{\xi^2+\nu_0^2}|x-\xi|}d\xi}{\Delta(\xi)}, \qquad (3.68)$$

where $M(\xi,y)$ and $\Delta(\xi)$ are the same as in Eq. (3.55), with $\mu_n = \sqrt{k_n^2 - \nu_0^2}$, k_n being roots of the dispersion relation in Eq. (2.26). Using the condition (3.67), it can be easily derived that

$$\frac{1}{2}\delta(y-\eta) = \sum_{n=0,I}^{II} i\mu_n A_n e^{-k_n y} - \frac{2}{\pi}\int_0^\infty \frac{\sqrt{\xi^2+\nu_0^2}A(\xi)M(\xi,y)d\xi}{\Delta(\xi)}. \qquad (3.69)$$

Using the orthogonal mode-coupling relation as defined in Eq. (2.30) for the functions $e^{-k_n y}$ and $M(\xi,y)$ as given in Eq. (3.69), the unknown constant A_n and the unknown function $A(\xi)$ are obtained as

$$A_n = -ie^{-k_n\eta}/\mu_n C_n, \ n=0,I,II, \quad A(\xi) = -\frac{M(\xi,\eta)}{2\sqrt{\xi^2+\nu_0^2}},$$

where C_n is the same as in Eq. (3.55). Thus, Green's function is expressed as

$$G(x,y;\xi,\eta) = -\sum_{n=0,I}^{II} \frac{i}{\mu_n C_n} e^{i\mu_n|x-\xi|-k_n(y+\eta)}$$
$$- \frac{1}{\pi}\int_0^\infty \frac{M(\xi,\eta)M(\xi,y)e^{-\sqrt{\xi^2+\nu_0^2}|x-\xi|}d\xi}{\sqrt{\xi^2+\nu_0^2}\,\Delta(\xi)}. \qquad (3.70)$$

Proceeding in a similar manner, Green's function $G(x,y;\xi,\eta)$ in water of finite depth is obtained as

$$G(x,y;\xi,\eta) = \sum_{n=0,I}^{II} A_n f_n(y)e^{i\mu_n|x-\xi|} + \sum_{n=1}^\infty B_n f_n(y)e^{-\sqrt{k_n^2+\nu_0^2}|x-\xi|}, \qquad (3.71)$$

where f_n is the same as in Eq. (3.64) with $k_n = ik_n$ for $n=1,2,....$ Applying the condition (3.67), Eq. (3.71) yields

$$\frac{1}{2}\delta(y-\eta) = \sum_{n=0,I}^{II} i\mu_n A_n f_n(y) - \sum_{n=1}^\infty \sqrt{k_n^2+\nu_0^2}B_n f_n(y).$$

Applying the orthogonal mode-coupling relation in Eq. (2.50), the unknown constants are obtained as

$$A_n = \frac{-if_n(\eta)}{2\mu_n C_n}, \ n=0,I,II, \quad B_n = \frac{-f_n(\eta)}{2C_n\sqrt{k_n^2+\nu_0^2}}, \ n=1,2,.... \qquad (3.72)$$

Substituting for A_n and B_n from Eq. (3.72) in Eq. (3.71), Green's function in the case of finite water depth is obtained explicitly as

$$G(x,y;\xi,\eta) = \frac{-i}{2} \sum_{n=0,I}^{II} \frac{f_n(\eta)f_n(y)}{\mu_n C_n} e^{i\mu_n|x-\xi|}$$

$$- \frac{1}{2} \sum_{n=1}^{\infty} \frac{f_n(\eta)f_n(y)}{C_n\sqrt{k_n^2+\nu_0^2}} e^{-\sqrt{k_n^2+\nu_0^2}|x-\xi|}. \tag{3.73}$$

An alternate form of Green's function for the above discussed problem can be derived using the splitting technique as discussed in Subsection 3.1 and details can be found in [26].

3.3.2   Expansion formulae based on Green's identity

For a large class of wave structure interaction problems, often the free surface Green's function is used. However, the utility of Green's function for the wave structure interaction problem will be demonstrated here using Green's identity to derive the expansion formulae for the wave structure interaction problem in the cases of both infinite and finite water depths. The associated BVPs are the same as described in Subsection 3.2.1 given by the Eqs. (3.39)–(3.42) and (3.44). In the present case, condition (3.43) is replaced by

$$\frac{\partial \phi}{\partial x} = u(y) \quad \text{on} \quad x=0, 0<y<h. \tag{3.74}$$

Consider

$$G^{mod}(x,y;\xi,\eta) = G(x,y;\xi,\eta) + G(-x,y;\xi,\eta), \tag{3.75}$$

with zero normal velocity on the wavemaker, that is, $G_x^{mod}(0,y;\xi,\eta) = 0$. Using Green's identity, the velocity potential $\phi(x,y)$ is obtained as

$$\phi(\xi,\eta) = \int_0^\infty G^{mod}(0,y;\xi,\eta)u(y)dy$$

$$+ \int_0^\infty \{G^{mod}(x,0;\xi,\eta)\phi_y(x,0) - \phi(x,0)G_y^{mod}(x,0;\xi,\eta)\}dx, \tag{3.76}$$

where the contribution from the bottom boundary and at infinity become zero. Using the fact that $G(x,y,\xi,\eta)$ is outgoing in nature at infinity and $G^{mod}(x,y;\xi,\eta)$ satisfies

$$G_{xy}^{mod}(0,0;\xi,\eta) = 2G_y(0,0;\xi,\eta), \quad G_{xy}^{mod}(0,0;\xi,\eta) = 0,$$

$$G_{yyy}^{mod}(0,0;\xi,\eta) = 2G_{yyy}(0,0;\xi,\eta), \quad G_{xxxy}^{mod}(0,0;\xi,\eta) = 0,$$

it can be easily derived that

$$\int_0^\infty \{G^{mod}(x,0;\xi,\eta)\phi_y(x,0) - \phi(x,0)G_y^{mod}(x,0;\xi,\eta)\}dx$$

$$= \frac{2}{K}\Big[QG_y\phi_{xy} + D\{G_{yyy}\phi_{xy} + G_y\phi_{xyyy}\}\Big]_{(x,y)=(0,0)}. \tag{3.77}$$

Using the fact that

$$G^{mod}(0, y; \xi, \eta) = 2G(0, y; \xi, \eta) = 2G(\xi, \eta; 0, y),$$

from Eqs. (3.75) and (3.76), the velocity potential is obtained as

$$
\begin{aligned}
\phi(\xi, \eta) &= 2 \int_{\Re} G(0, y; \xi, \eta) u(y) dy + \frac{2}{K} \Big[Q G_y(0, 0; \xi, \eta) \phi_{xy}(0, 0) \\
&+ D\{ G_{yyy}(0, 0; \xi, \eta) \phi_{xy}(0, 0) + G_y(0, 0; \xi, \eta) \phi_{xyyy}(0, 0) \} \Big], \quad (3.78)
\end{aligned}
$$

where $\Re$ is the semi-infinite interval $(0, \infty)$ in the case of infinite water depth and the finite interval $(0, h)$ in the case of finite water depth.

In the case of infinite water depth, substituting the explicit forms of Green's function from Eq. (3.70) into Eq. (3.78), the expansion formula for the velocity potential is obtained as

$$
\phi(x, y) = \sum_{n=0, I}^{II} A_n e^{i\mu_n x - k_n y} + \int_0^{\infty} \frac{a(t) e^{-x \sqrt{t^2 + \nu_0^2}} M(t, y) dt}{\sqrt{t^2 + \nu_0^2} \Delta(t)}, \quad (3.79)
$$

where

$$
\begin{aligned}
a(t) &= -\frac{1}{\pi} \int_0^{\infty} M(t, y) u(y) dy \\
&- \xi \Big[D\{ -\xi^2 \phi_y + \phi_{yyy} \} + Q\phi_y \Big]_{(x, y) = (0, 0)}, \\
A_n &= -\frac{i}{\mu_n C_n} \Big[\int_0^{\infty} e^{-k_n y} u(y) dy \\
&- \frac{k_n}{K} \{ D(k_n^2 \phi_y + \phi_{yyy}) - Q\phi_y \}_{(x, y) = (0, 0)} \Big], \quad n = 0, I, II.
\end{aligned}
$$

Here, all the unknowns are expressed in terms of the two unknowns $\phi_y(0, 0)$ and $\phi_{yyy}(0, 0)$. These two quantities have to be determined from the proposed edge conditions of the physical problem.

On the other hand, in the case of finite water depth, using the explicit form of $G(x, y; \xi, \eta)$ from Eq. (3.73) and proceeding in a similar manner as in the case of infinite water depth, Eq. (3.78) yields

$$
\phi(x, y) = \sum_{n=0, I}^{II} A_n f_n(y) e^{i\mu_n x} + \sum_{n=1}^{\infty} a_n f_n(y) e^{-\sqrt{k_n^2 + \nu_0^2} x}, \quad (3.80)
$$

where

$$A_n = \frac{i}{C_n \mu_n} \left[\int_0^h u(t) f_n(t) dt + \frac{k_n \tanh k_n h}{K} \left\{ Q\phi_{xy} \right. \right.$$

$$\left. \left. + D(k_n^2 \phi_{xy} + \phi_{xyyy}) \right\}_{(x,y)=(0,0)} \right], \quad n = 0, I, II,$$

$$(3.81)$$

$$a_n = -\frac{1}{\sqrt{p_n^2 + \nu_0^2} \, C_n} \left[\int_0^h u(t) f_n(t) dt + \frac{k_n \tan k_n h}{K} \left\{ Q\phi_y \right. \right.$$

$$\left. \left. + D(-k_n^2 \phi_y + \phi_{yyy}) \right\}_{(x,y)=(0,0)} \right], \quad n = 1, 2, ...$$

$$(3.82)$$

Thus, Eq. (3.80) gives the expansion formula for the velocity potential in finite water depth.

Proceeding in a similar manner as discussed above, choosing $G^{mod}(x, y; \xi, \eta)$ $= G(x, y; \xi, \eta) - G(-x, y; \xi, \eta)$, the velocity potential associated with the boundary value problem given by Eqs. (3.39)–(3.44) can be derived by the suitable application of Green's identity and details are left as an exercise. Further, for $\nu_0 = 0$, these formulae agree with results associated with the two-dimensional Laplace equation as derived in [87]. Further, proceeding in a similar manner as discussed in Subsection 3.2, the expansion formulae associated with the boundary value problem given by Eqs. (3.39)–(3.42), (3.44) and (3.74) can be obtained with the help of the Fourier cosine transform along the x-axis in the cases of water of both finite and infinite depths.

3.3.3 Fundamental singularities in two-layer fluids

In this subsection, first Green's function associated with wave structure interaction problems in two-layer fluids in three dimensions is derived under the assumption that the velocity potential is harmonic along the z-axis and is of the form $\Phi(x, y, z, t) = Re\{\phi(x, y)e^{i\nu_0 z - i\omega t}\}$. Further, a review is provided on various types of singularities at the end of the subsection. The formulation of the general problem is discussed in Subsection 1.4.4. Thus, Green's function $G(x, y; \xi, \eta)$ associated with the wave structure interaction problems will satisfy

$$\nabla^2 G - \nu_0^2 G = 0, \qquad (3.83)$$

in the fluid domain except the point (ξ, η). The bottom boundary condition is given by

$$\frac{\partial G}{\partial y} = 0 \quad \text{at} \quad y = H \quad \text{in the case of finite depth} \qquad (3.84)$$

and

$$G, |\nabla G| \to 0 \quad \text{as} \quad y \to \infty \quad \text{in the case of infinite depth.} \qquad (3.85)$$

The boundary condition at the mean free surface at $y = 0$ on the floating flexible plate is given by (as in Subsection 1.4.4)

$$D_1 \frac{\partial^5 G}{\partial y^5} - Q_1 \frac{\partial^3 G}{\partial y^3} + \frac{\partial G}{\partial y} + KG = 0 \text{ on } y = 0,\ 0 \le x < \infty, \qquad (3.86)$$

and the conditions at the mean interface on the flexible submerged plate surface for $0 < x < \infty$ are given by

$$\left. \frac{\partial G}{\partial y} \right|_{y=h+} = \left. \frac{\partial G}{\partial y} \right|_{y=h-} \qquad (3.87)$$

and

$$\left(D_2 \frac{\partial^5 G}{\partial y^5} - Q_2 \frac{\partial^3 G}{\partial y^3} + \frac{\partial G}{\partial y} + KG \right)\bigg|_{y=h+} = s \left(\frac{\partial G}{\partial y} + KG \right)\bigg|_{y=h-}. \qquad (3.88)$$

In Eqs. (2.136) and (2.138), D_1, D_2, Q_1, Q_2, K and s are known positive constants as in Subsection 1.4.4. In addition, $G(x, y; \xi, \eta)$ satisfies the condition

$$G(x, y; \xi, \eta) \sim K_0(\nu_0 r), \quad \text{as} \quad r = \sqrt{(x - \xi)^2 + (y - \eta)^2} \to 0, \qquad (3.89)$$

where $K_0(\nu_0 r)$ is the zeroth order modified Bessel function. Proceeding in a similar manner as discussed in Subsection 3.3.1, it can be easily derived that

$$\lim_{x \to 0} \frac{\partial G(x, y; 0, \eta)}{\partial x} = 2\pi \delta(y - \eta), \qquad (3.90)$$

where $\delta(y)$ is the delta function as defined in Eq. (3.1). The radiation condition in this case is of the form

$$G \to \begin{cases} \displaystyle\sum_{n=I}^{II} A_n F_n(k_n, y) e^{i\mu_n |x - \xi|}, \text{ as } |x - \xi| \to \infty, \text{(infinite depth)}, \\[4mm] \displaystyle\sum_{n=I}^{II} B_n \psi_n(k_n, y) e^{i\mu_n |x - \xi|}, \text{ as } |x - \xi| \to \infty, \text{(finite depth)}, \end{cases} \qquad (3.91)$$

where A_n and B_n are constants to be determined and k_n satisfies the dispersion relation Eq. (2.140) in the case of infinite water depth (Eq. (2.149) in the case of finite water depth) with $k_n^2 = \nu_0^2 + \mu_n^2$ for $n = 0, I, II$. Further, $F_n(k_n, y)$ and $\psi_n(k_n, y)$ are the same as in Eq. (2.139) and (2.247) in the case of water of infinite and finite depths, respectively. Before proceeding further, one of the identities associated with the delta function is given by (as in [100])

$$\int_0^\infty \delta(y - \eta) F(y) dy = \begin{cases} F(\eta) & \text{if } \eta > 0, \\ F(\eta)/2 & \text{if } \eta = 0. \end{cases} \qquad (3.92)$$

Green's function $G(x,y;\xi,\eta)$ satisfying Eq. (3.83), along with the boundary conditions Eqs. (3.85)–(3.90) in water of infinite depth, is given by

$$G(x,y;\xi,\eta) = \sum_{n=I}^{X} A_n F_n(k_n,y)e^{i\mu_n|x-\xi|} + \int_0^{\infty} A(\xi)L(\xi,y)e^{-\sqrt{\xi^2+\nu_0^2}|x-\xi|}d\xi,$$

(3.93)

for $x - \xi \neq 0$ with $F_n(k_n,y)$s and $L(\xi,y)$ being the same as defined in Eq. (2.139) with μ_n and ν_n being given by $k_n^2 = \nu_0^2 + \mu_n^2$ for $n = I, II, ..., X$. Keeping the realistic nature of physical problems in mind, it is assumed that the dispersion relation in Eq. (2.140) has two distinct real positive roots μ_I and μ_{II} and eight complex roots $\mu_{III}, \mu_{IV}, ..., \mu_X$ of the forms $a \pm ib$ and $-c \pm id$ as in Chapter 2. Thus, the boundedness criteria of the velocity potential $\phi(x,y)$ at the far field yields $A_{VII} = ... = A_X = 0$. Further, using the orthogonal mode-coupling relation as in Eq. (2.141) and the known identity in Eq. (3.92), from Eqs. (3.90) and (3.93), the unknowns $A_n, n = I, ..., VI, A(\xi)$ associated with Green's function $G(x,y;\xi,\eta)$ are obtained as

$$A_n = \begin{cases} \dfrac{-is\delta_1 f_n(\eta)}{2\mu_n C_n}, & 0 \leq \eta < h, \\[2ex] \dfrac{-i\{sf_n(h)+1\}}{2\mu_n C_n}, & \eta = h, \\[2ex] \dfrac{-is\delta_1 e^{-\mu_n(\eta-h)}}{\mu_n C_n}, & h < \eta < \infty, \end{cases}$$

$$A(\xi) = \begin{cases} \dfrac{-\delta_1 s K\{\xi T_1(\xi,2)\cos\xi\eta - K\sin\xi\eta\}}{\pi\sqrt{\xi^2+\nu_0^2}\Delta(\xi)}, & 0 \leq \eta < h, \\[2ex] \dfrac{-[(1+s)K\{\xi T_1(\xi,2)\cos\xi h - K\sin\xi h\} - W(\xi)]}{\pi\sqrt{\xi^2+\nu_0^2}\Delta(\xi)}, & \eta = h, \\[2ex] \dfrac{-[K\{\xi T_1(\xi,2)\cos\xi\eta - K\sin\xi\eta\} - W(\xi)\cos\xi(\eta-h)]}{\pi\sqrt{\xi^2+\nu_0^2}\Delta(\xi)}, & \eta > 0, \end{cases}$$

with μ_n $F_n(y)$, $L(\xi,y)$, $f_n(y)$, $W(\xi)$, C_n and $\Delta(\xi)$ being the same as given in Subsection 2.6.2 and

$$\delta_1 = \begin{cases} 1 & \text{for} \quad 0 < \eta < h, \\ 1/2 & \text{for} \quad \eta = 0. \end{cases}$$

Proceeding in a similar manner as above, in the case of finite water depth Green's function $G(x,y;\xi,\eta)$ is obtained as

$$G(x,y;\xi,\eta) = \sum_{n=I}^{VI} A_n \psi_n(y)e^{i\mu_n(x-\xi)} + \sum_{n=1}^{\infty} B_n \psi_n(y)e^{-\sqrt{k_n^2+\nu_0^2}(x-\xi)}, \quad (3.94)$$

where A_n is given by

$$
A_n = \begin{cases}
\dfrac{-s\delta_2 \sinh k_n(H-h)L_1(ik_n;\eta)}{2\mu_n C_n K \mathcal{D}(ik_n,h)}, & \text{for } 0 \le \eta < h, \\[3mm]
\dfrac{-s\sinh k_n(H-h)L_1(ik_n;h)}{2\mu_n C_n} - \dfrac{i\cosh k_n(H-h)}{2\mu_n C_n}, & \text{for } \eta = h, \\[3mm]
\dfrac{-i\delta_2 \cosh k_n(H-\eta)}{2\mu_n C_n}, & \text{for } h < \eta \le H,
\end{cases}
$$

and B_n, $n = 1, 2, 3, \ldots$ is obtained from the expression for A_n with $k_n = ik_n$, where $\delta_2 = 1$ for $\eta \in (0, h) \cup (h, H)$ and $\delta_2 = 1/2$ for $\eta = 0, H$ and k_n, $\psi_n(y)$, $L_1(ik_n; \eta)$, C_ns being the same as in Subsection 2.6.3.

As an application of the fundamental line source potential, multipole line source potentials are derived next in the cases of both infinite and finite depths. The BVP for symmetric and antisymmetric multipole potentials $G_s(x, y; \xi, \eta)$ and $G_a(x, y; \xi, \eta)$ remain the same as in the case of fundamental source potentials except for the condition (3.89), which is replaced by

$$
G_s \sim \frac{1}{2\pi}\frac{\cos n\theta}{r^n}, \quad G_a \sim \frac{1}{2\pi}\frac{\sin n\theta}{r^n}, \quad \text{as} \quad r \to 0 \quad \text{for} \quad n = 1, 2, \ldots, \quad (3.95)
$$

where $x - \xi = r\cos\theta$, $y - \eta = r\sin\theta$ with $0 < \theta < \pi$. Using the following known representations ([82])

$$
\log r = \int_0^\infty \frac{1}{\xi}\left(e^{-\xi} - e^{-\xi|y-\eta|}\right)\cos\xi(x-\xi)d\xi, \quad y, \eta > 0,
$$

$$
\frac{\cos n\theta}{r^n} = \begin{cases}
\dfrac{1}{(n-1)!}\displaystyle\int_0^\infty \xi^{n-1}e^{-\xi(y-\eta)}\cos\xi(x-\xi)d\xi, & y > \eta, \\[3mm]
\dfrac{(-1)^n}{(n-1)!}\displaystyle\int_0^\infty \xi^{n-1}e^{-\xi(\eta-y)}\cos\xi(x-\xi)d\xi, & y < \eta,
\end{cases}
$$

$$
\frac{\sin n\theta}{r^n} = \begin{cases}
\dfrac{1}{(n-1)!}\displaystyle\int_0^\infty \xi^{n-1}e^{-\xi(y-\eta)}\sin\xi(x-\xi)d\xi, & y > \eta, \\[3mm]
\dfrac{(-1)^{n+1}}{(n-1)!}\displaystyle\int_0^\infty \xi^{n-1}e^{-\xi(\eta-y)}\cos\xi(x-\xi)d\xi, & y < \eta,
\end{cases}
$$

it can be derived that

$$
\frac{\cos n\theta}{r^n} = -\frac{1}{(n-1)!}\frac{\partial^n \log r}{\partial \eta^n},
$$

$$
\frac{\sin n\theta}{r^n} = -\frac{1}{(n-1)!}\frac{\partial^n \log r}{\partial \xi \partial \eta^{n-1}}.
$$

Therefore, the expansions for the symmetric and antisymmetric multipole line source potentials can be represented as

$$G_s(x, y; \xi, \eta) = -\frac{1}{(n-1)!} \frac{\partial^n U}{\partial \eta^n}, \quad n = 1, 2, ..., \tag{3.96}$$

$$G_a(x, y; \xi, \eta) = -\frac{1}{(n-1)!} \frac{\partial^n U}{\partial \xi \partial \eta^{n-1}}, \quad n = 1, 2, ..., \tag{3.97}$$

where U is the fundamental wave source potential associated with the Laplace equation in two dimensions. The process is similar in the case of the reduced wave equation. In the case of reduced wave potential, the multipole behaves like the BVP for symmetric and antisymmetric multipole potentials. $G_s(x, y; \xi, \eta)$ and $G_a(x, y; \xi, \eta)$ remain the same as in the case of fundamental source potentials except for the condition (3.89), which is replaced by (as in [26])

$$G_s \sim \cos n\theta K_n(\nu_0 r), \quad G_a \sim \sin n\theta K_n(\nu_0 r), \quad \text{as} \quad r \to 0 \quad \text{for} \quad n = 1, 2, ..., \tag{3.98}$$

where $x - \xi = r \cos \theta$, $y - \eta = r \sin \theta$ with $0 < \theta < \pi$. From the asymptotic behavior of $K_n(\nu_0 r)$, it can be easily derived that G_s and G_a behave in the same manner as in Eq. (3.96) and (3.97) from which the total potential in the presence of multipoles can be obtained in a straightforward manner as in [88].

Further details on fundamental singularities derived for a wide class of physical problems associated with surface gravity waves can be found in [100], [108], [117], [142] and the literature cited therein. In the context of the results derived in the present chapter, source and multipole potentials can be computed for problems in cylindrical and polar coordinates to deal with wave diffraction by circular flexible plates and discs, some of which are left as exercises at the end this chapter. Expansion formulae for three-dimensional problems can be attempted for solutions based on the application of Green's function technique and Green's identity which will require the construction of appropriate Green's function. Certain results on three-dimensional wave structure interaction problems can be found in [104] and [103].

3.4   Scattering of surface waves by floating elastic plates

The utility of the expansion formulae is illustrated in this section for wave interactions with floating and submerged horizontal flexible plates. Two problems are discussed in detail.

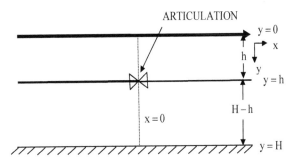

FIGURE 3.1
Schematic diagram of floating and articulated submerged plates.

3.4.1 Flexural gravity wave diffraction by an articulated submerged plate

In this subsection, the problem of structural response attenuation of a floating elastic plate by an articulated submerged elastic plate is studied in water of finite and infinite depths. The fluid and structural characteristics remain the same as discussed in Chapter 1, and the problem is analysed in the two-dimensional Cartesian coordinate system. The fluid occupies the region $-\infty < x < \infty, 0 < y < H$ in the case of finite water depth ($-\infty < x < \infty, 0 < y < \infty$ in the case of infinite water depth) except for the region occupied by the floating plate and the articulated submerged plate as in Figure 3.1. In order to obtain the velocity potential and analyse the effect of articulation on the submerged plate, the fluid domain is divided into two subdomains, namely, region 1 for $x > 0$ and region 2 for $x < 0$ in the cases of water of both finite and infinite depths. The submerged plate is assumed to be infinitely large and is a combination of two semi-infinitely large plates which are kept in the planes $x > 0$ and $x < 0$ at $y = h$. The velocity potentials $\phi_j(x, y)$, $j = 1, 2$ satisfy the two-dimensional Laplace equation as given by

$$\nabla^2 \phi_j = 0, \quad \text{in the fluid domain.} \tag{3.99}$$

The bottom boundary conditions are given by

$$\frac{\partial \phi_2}{\partial y} = 0 \quad \text{on} \quad y = H \quad \text{in the case of finite depth} \tag{3.100}$$

and

$$\phi, |\nabla \phi| \to 0 \quad \text{as} \quad y \to \infty \quad \text{in the case of infinite depth.} \tag{3.101}$$

The linearised boundary condition on the plate-covered surface at $y = 0$ is given by (as in Section 1.5.3)

$$D_1 \frac{\partial^5 \phi_1}{\partial y^5} - Q_1 \frac{\partial^3 \phi_1}{\partial y^3} + \frac{\partial \phi_1}{\partial y} + K\phi_1 = 0 \text{ on } y = 0, \ x \in (-\infty, \infty), \quad (3.102)$$

while, on the submerged articulated flexible plate at $y = h$, ϕ_1 and ϕ_2 satisfy

$$D_2 \frac{\partial^5 \phi_2}{\partial y^5} - Q_2 \frac{\partial^3 \phi_2}{\partial y^3} + K(\phi_2 - \phi_1) = 0, \text{ for } y = h, \ x \in (-\infty, \infty) \backslash \{0\}, \quad (3.103)$$

where $D_i = E_i I_i / \rho g$, $Q_i = N_i / \rho g$ and $K = w^2/g$. In addition, at the interface $x = 0\pm$, the continuity of pressure and velocity yields

$$\phi_1(0+, y) = \phi_2(0-, y) \quad (3.104)$$

$$\text{and} \quad \phi_{1x}(0+, y) = \phi_{2x}(0-, y). \quad (3.105)$$

The plates are connected at $x = 0\pm$, $y = h$ by a vertical linear spring with stiffness k_{33} and a flexural rotational spring with stiffness k_{55}. Thus, the shear force and bending moment at the connected edges $(0\pm, h)$ satisfy the following conditions in ϕ as given by (as in [62])

$$D_2 \phi_{1yxx}(0+, h) = k_{55}\{\phi_{1yx}(0+, h) - \phi_{2yx}(0-, h)\}, \quad (3.106)$$

$$D_2 \phi_{2yxx}(0-, h) = k_{55}\{\phi_{1yx}(0+, h) - \phi_{2yx}(0-, h)\}, \quad (3.107)$$

$$D_2 \phi_{1yxxx}(0+, h) = -k_{33}\{\phi_{1y}(x+, h) - \phi_{2y}(0-, h)\}, \quad (3.108)$$

$$D_2 \phi_{2yxxx}(0-, h) = -k_{33}\{\phi_{1y}(x+, h) - \phi_{2y}(0-, h)\}. \quad (3.109)$$

At the interface, the deflection, slope of deflection, bending moment and shear force acting on the floating elastic plate are assumed to be continuous at $(0,0)$ and $(\pm a, 0)$ (as in [61]), which yields

$$\frac{\partial \phi}{\partial y}\bigg|_{(d+,0)} = \frac{\partial \phi}{\partial y}\bigg|_{(d-,0)}, \qquad \frac{\partial^2 \phi}{\partial x \partial y}\bigg|_{(d+,0)} = \frac{\partial^2 \phi}{\partial x \partial y}\bigg|_{(d-,0)}, \quad (3.110)$$

$$\frac{\partial^3 \phi}{\partial y \partial x^2}\bigg|_{(d+,0)} = \frac{\partial^3 \phi}{\partial y \partial x^2}\bigg|_{(d-,0)}, \qquad \frac{\partial^4 \phi}{\partial y \partial x^3}\bigg|_{(d+,0)} = \frac{\partial^4 \phi}{\partial y \partial x^3}\bigg|_{(d-,0)}, \quad (3.111)$$

for $d = 0, \pm a$. Further, the far field radiation condition is given by

$$\phi(x, y) = \begin{cases} \sum\limits_{n=I}^{II} (C_n e^{-ip_n x} + R_n e^{ip_n x}) g_n(y) & \text{as} \quad x \to \infty, \\[4mm] \sum\limits_{n=I}^{II} T_n e^{-ip_n x} g_n(y) & \text{as} \quad x \to -\infty, \end{cases} \quad (3.112)$$

where C_i is the amplitude of the incident waves assumed to be known, R_n and T_n are the unknown constants associated with the amplitude of the reflected and transmitted waves and are to be determined. Here, subscripts I and II refer to the waves generated due to the interaction of surface gravity waves with the floating and submerged plates, respectively. Further, $g_n(y)$ is defined as

$$g_n(y) = \begin{cases} F_n(y), & \text{in the case of infinite water depth,} \\ \psi_n(y), & \text{in the case of finite water depth,} \end{cases} \tag{3.113}$$

where

$$F_n(y) = \begin{cases} \dfrac{-iL_1(i\mu_n, y)}{KD(i\mu_n, h)}, & 0 < y < h, \\ e^{-\mu_n(y-h)}, & h < y < \infty, \end{cases}$$

$$\psi_n(y) = \begin{cases} -\dfrac{i\tanh p_n(H-h)L_1(ip_n, y)}{KD(ip_n, h)}, & 0 < y < h, \\ \dfrac{\cosh p_n(H-y)}{\cosh p_n(H-h)}, & h < y < H, \end{cases}$$

$$L_1(\xi, y) = K\{T_1(\xi, 2)\xi \cos \xi y - K \sin \xi y\},$$
$$D(\xi, h) = T_1(\xi, 2)\xi \sin \xi h + K \cos \xi h,$$
$$T_1(\xi, 2) = D_1\xi^4 + Q_1\xi^2 + 1$$

and p_n satisfies the dispersion relation

$$\mathcal{G}(p_n) \equiv \frac{(D_2 p_n^4 - Q_2 p_n^2)p_n - K \coth p_n(H-h)}{iL_1(ip_n, h)} - \frac{1}{D(ip_n, h)} = 0, \tag{3.114}$$

in the case of finite water depth (in the case of infinite water depth with $H \to \infty$). Exploiting the geometrical symmetry of the physical problem about $x = 0$, the boundary value problem in ϕ is split into the reduced potentials defined by (as in [33])

$$\varphi(x, y) = \phi_1(x, y) - \phi_2(-x, y) \text{ and } \Upsilon(x, y) = \phi_1(x, y) + \phi_2(-x, y). \tag{3.115}$$

Thus, the reduced potentials $\varphi(x, y)$ and $\Upsilon(x, y)$ will satisfy the governing Eq. (3.99) in the region $0 < x < \infty$ along with the plate-covered boundary conditions in Eqs. (3.102) and (3.103), and the bottom conditions as in Eqs. (3.100) and (3.101). The edge conditions in Eqs. (3.106)–(3.109) for the bending moment and shear force in terms of $\varphi(x, y)$ and $\Upsilon(x, y)$ are given by

$$D_2 \partial_{xy^3}^4 \varphi(0, h) = 2k_{33}\partial_y \varphi(0, h), \quad \partial_y^3 \varphi(0, h) = 0, \tag{3.116}$$

$$D_2 \partial_y^3 \Upsilon(0, h) = -2k_{55}\partial_{xy}^2 \Upsilon(0, h), \quad \partial_{xy^3}^4 \Upsilon(0, h) = 0. \tag{3.117}$$

The continuity of deflection, slope of deflection, bending moment and shear force acting on the floating elastic plate in Eqs. (3.110) and (3.111) yield

$$\partial_y \varphi(0, 0) = 0, \quad \partial_{yxx}^3 \varphi(0, 0) = 0, \tag{3.118}$$

$$\partial_{yx}^2 \Upsilon(0,0) = 0, \quad \partial_{yxxx}^4 \Upsilon(0,0) = 0. \tag{3.119}$$

The far field condition in Eq. (3.112) yields

$$\varphi(x,y) \sim \sum_{n=I}^{II} \{C_n e^{-ip_n x} + A_n e^{ip_n x}\} g_n(y), \quad \text{as} \quad x \to \infty, \tag{3.120}$$

$$\Upsilon(x,y) \sim \sum_{n=I}^{II} \{C_n e^{-ip_n x} + B_n e^{ip_n x}\} g_n(y), \quad \text{as} \quad x \to \infty, \tag{3.121}$$

where $A_n = R_n - T_n$ and $B_n = R_n + T_n$. Finally, the continuity of pressure and velocity as in Eqs. (3.104) and (3.105) yields

$$\varphi(0,y) = 0 \text{ and } \Upsilon_x(0,y) = 0, \text{ at } x = 0. \tag{3.122}$$

Thus, the boundary value problems in terms of the reduced potentials $\varphi(x,y)$ and $\Upsilon_x(x,y)$ are defined in a quarter plane $(0 < x < \infty, 0 < y < \infty)$ in the case of water of infinite depth and semi-infinite strip $(0 < x < \infty, 0 < y < h)$ in the case of water of finite depth. Next, using the expansion formulae as discussed in Section 3.2 (similar to the expansion formulae discussed in Section 2.6 in the case of homogeneous fluid), the reduced potentials are derived in a straightforward manner from which the total potential and the deflection of the plates and other physical quantities can be easily derived.

In the case of infinite water depth, using the expansion formulae as in Section 2.6 in the case of homogeneous fluid with the reduced velocity potential $\varphi(x,y)$, satisfying Eq. (3.99), the boundary conditions in Eqs. (3.101)–(3.103) are obtained as

$$\varphi(x,y) = \sum_{n=I}^{II} C_n F_n(y) e^{-i\mu_n x} + \sum_{n=I}^{VI} A_n F_n(y) e^{i\mu_n x} + \int_0^\infty A(\xi) L(\xi, y) e^{-\xi x} d\xi, \tag{3.123}$$

where $F_n(y)$ is the same as defined in Eq. (3.113) and $L(\xi, y)$ is given by

$$L(\xi, y) = \begin{cases} L_1(\xi, y) & \text{for } 0 < y < h, \\ L_2(\xi, y) & \text{for } h < y < \infty, \end{cases}$$

$$L_2(\xi, y) = L_1(\xi, y) - \{-\xi(D_2\xi^4 + Q_2\xi^2)\mathcal{D}(\xi, h)\} \cos \xi(y - h), \tag{3.124}$$

with $L_1(\xi, y)$ and $\mathcal{D}(\xi, h)$ being the same as defined in Eq. (3.113). The eigenfunctions F_n and $L(\xi, y)$ defined in Eqs. (3.113) and (3.123) satisfy the orthogonal mode-coupling relation given by (see Subsection 2.6.3)

$$\langle F_m, F_n \rangle = \langle F_m, F_n \rangle_1 + \langle F_m, F_n \rangle_2 = C_n \delta_{mn}, \quad m, n = I, ..., VI, \tag{3.125}$$

$$\langle F_n, L(\xi, y) \rangle = 0 \quad \text{for } \xi > 0, \ n = I, ..., VI, \tag{3.126}$$

where

$$\langle F_m, F_n \rangle_1 = \int_0^h F_m(y) F_n(y) dy - \frac{Q_1}{K} F_m'(0) F_n'(0)$$
$$+ \frac{D_1}{K} \left\{ F_m'(0) F_n'''(0) + F_m'''(0) F_n'(0) \right\},$$

$$\langle F_m, F_n \rangle_2 = \lim_{\epsilon \to 0} \int_h^\infty e^{-\epsilon y} F_m(y) F_n(y) dy - \frac{Q_2}{K} F_m'(h) F_n'(h)$$
$$+ \frac{D_2}{K} \left\{ F_m'(h) F_n'''(h) + F_m'''(h) F_n'(h) \right\},$$

$$C_n = -\mathcal{D}(i\mu_n h) \sinh \mu_n h [\psi_n'(0)]^2 \mathcal{G}'(\mu_n) / 2K^2 \mu_n^2$$

and

$$\mathcal{G}(\mu_n) = K - \frac{\mu_n T_1(i\mu_n, 2)\{K(\coth \mu_n h + 1) - \mu_n(D_2 \mu_n^4 - Q_2 \mu_n^2)\}}{K(1 + \coth \mu_n h) - (D_2 \mu_n^4 - Q_2 \mu_n^2)\mu_n \coth \mu_n h}, \quad (3.127)$$

with $\mathcal{D}(i\mu_n, h) \neq 0$. In Eq. (3.127), $\mathcal{G}(\mu_n) = 0$ yields the dispersion relation satisfied by μ_n in Eq. (3.123). Using the equation of continuity as in Eq. (3.121) along with the edge conditions as in Eqs. (3.116) and (3.118) and the orthogonal mode-coupling relation in Eq. (3.125) from Eq. (3.123), the unknowns A_n for $n = I, ..., VI$ and $A(\xi)$ are obtained as

$$A_n = -\frac{\alpha_1(D_2 \mu_n^3 - Q_2 \mu_n)}{C_n K} - C_n, \quad A(\xi) = \frac{2\alpha_1(D_2 \xi^3 + Q_2 \xi)\mathcal{D}(\xi, h)}{\pi \triangle(\xi)}, \quad (3.128)$$

where $\triangle(\xi) = W^2(\xi) - 2KW(\xi)L_1(\xi, h) + \{\xi^2 T_1^2(\xi, 2) + K^2\}K^2,$

$$\alpha_1 = 2 \sum_{n=I}^{II} iD_2 \mu_n^4 C_n / \beta_1,$$

$$\beta_1 = \int_0^\infty \frac{2K(D_2 \xi^4 - 2k_{33}\xi)(D_2 \xi^3 + Q_2 \xi)\mathcal{D}^2(\xi, h) d\xi}{\pi \triangle(\xi)}$$
$$- \sum_{n=I}^{VI} \frac{(D_2 \mu_n^4 - Q_2 \mu_n)(iD_2 \mu_n^4 - 2k_{33}\mu_n)}{K C_n},$$

with C_ns being the same as defined in Eq. (3.125). Proceeding in a similar manner as in the case of $\varphi(x, y)$, the reduced velocity potential $\Upsilon(x, y)$ satisfying the governing Eq. (3.99) and boundary conditions in Eqs. (3.101)–(3.103) in the case of infinite water depth is obtained as

$$\Upsilon(x, y) = \sum_{n=I}^{II} C_n F_n(y) e^{-i\mu_n x} + \sum_{n=I}^{VI} B_n F_n(y) e^{i\mu_n x} + \int_0^\infty B(\xi) L(\xi, y) e^{-\xi x} d\xi.$$

$$(3.129)$$

Using the continuity condition in Eq. (3.122), edge conditions as in Eqs. (3.117) and (3.119) and the orthogonal mode-coupling relation as in Eqs. (3.125)–(3.126), the unknown coefficients B_n for $n = I, ..., VI, B(\xi)$ are obtained as

$$B_n = -\frac{\alpha_2(D_2\mu_n^3 - Q_2\mu_n)}{i\mu_n C_n K} + C_n,$$

$$B(\xi) = \frac{2\alpha_2(D_2\xi^3 + Q_2\xi)\mathcal{D}(\xi,h)}{\pi\xi\triangle(\xi)}$$

$$\alpha_2 = 2\sum_{n=I}^{II} D_2\mu_n^3 C_n/\beta_2,$$

$$\beta_2 = \sum_{n=I}^{VI} \frac{(D_2\mu_n^3 - Q_2\mu_n)(D_2\mu_n^2 + 2k_{55}i\mu_n)}{iKC_n}$$

$$- \frac{2K}{\pi}\int_0^\infty \frac{(D_2\xi + 2k_{55})(D_2\xi^4 + Q_2\xi^2)\mathcal{D}^2(\xi,h)d\xi}{\triangle(\xi)}.$$

Proceeding in a similar manner as in the case of infinite water depth, using the expansion formulae as in Section 2.6 for homogeneous fluid in the case of finite water depth, the reduced velocity potential $\varphi(x,y)$ satisfying the governing Eq. (3.99) and the boundary conditions in Eqs. (3.100), (3.102) and (3.103) is obtained as

$$\varphi(x,y) = \sum_{n=I}^{II} C_n\psi_n(y)e^{-ip_nx} + \sum_{n=I,...VI,1}^{\infty} A_n\psi_n(y)e^{ip_nx}, \qquad (3.130)$$

where $\psi_n(y)$ is the same as defined in Eq. (3.113) with k_n satisfying the dispersion relation (3.114). It is assumed that k_I, k_{II} are real, $k_{III}, ..., k_{VI}$ are complex roots leading to the bounded solution, $k_{VII}, ..., k_x$ are complex roots of Eq. (3.114) leading to the unbounded solution and are not considered in Eq. (3.130) and k_n for $n = 1, 2, ...$ are the imaginary roots of Eq. (3.114) leading to the evanescent modes in Eq. (3.130). Further, the eigenfunctions ψ_n satisfy the orthogonal mode-coupling relation as given by

$$\langle\psi_m, \psi_n\rangle = \langle\psi_m, \psi_n\rangle_3 + \langle\psi_m, \psi_n\rangle_4 = C_n\delta_{mn}, \quad m, n = I, ...VI, 1, 2, ..., \quad (3.131)$$

$$\text{with } \langle\psi_m, \psi_n\rangle_3 = \int_0^h \psi_m(y)\psi_n(y)dy$$

$$- \frac{Q_1}{K}\psi'_m(0)\psi'_n(0) + \frac{D_1}{K}[\psi'_m(0)\psi'''_n(0) + \psi'''_m(0)\psi'_n(0)],$$

$$\langle \psi_m, \psi_n \rangle_4 = \int_h^H \psi_m(y)\psi_n(y)dy$$
$$- \frac{Q_2}{K}\psi_m'(h)\psi_n'(h) + \frac{D_2}{K}[\psi_m'(h)\psi_n'''(h) + \psi_m'''(h)\psi_n'(h)],$$

$$C_n = \frac{-\mathcal{D}(ip_n, h)\sinh p_n h[\psi_n'(0)]^2 \mathcal{G}'(p_n)}{2K^2 p_n^2},$$

where $\mathcal{G}(p_n)$ and $\mathcal{D}(p_n, h)$ are the same as defined in Eqs. (3.114) and (3.113), respectively. Using the equation of continuity as in Eq. (3.122) along with the edge conditions as in Eqs. (3.116) and (3.118) and the orthogonal mode-coupling relation in Eq. (3.131), from Eq. (3.130) the unknown coefficients A_n for $n = I, ..., VI, 1, 2, ...$ are obtained as

$$A_n = \frac{\alpha_1(D_2 p_n^3 - Q_2 p_n)\sinh p_n(H - h)}{C_n K} - C_n, \qquad (3.132)$$

with $\quad \alpha_1 = 2i \sum_{n=I}^{II} D_2 p_n^4 C_n \sinh p_n(H - h)/\beta_1,$

$$\beta_1 = \sum_{n=I,...,VI,1}^{\infty} \frac{(D_2 i p_n^4 - 2k_{33}p_n)(D_2 p_n^3 - Q_2 p_n)\sinh^2 p_n(H - h)}{KC_n}.$$

Proceeding in a similar way, the reduced velocity potential $\Upsilon(x, y)$ satisfying the governing Eq. (3.99) and the boundary conditions in Eqs. (3.100), (3.102) and (3.103) is obtained as

$$\Upsilon(x, y) = \sum_{n=I}^{II} C_n \psi_n(y)e^{-ip_n x} + \sum_{n=I,...,VI,1}^{\infty} B_n \psi_n(y)e^{ip_n x}, \qquad (3.133)$$

where C_n is the same as defined in Eq. (3.112). Using the continuity condition in Eq. (3.122), the edge conditions as in Eqs. (3.117) and (3.119) and the orthogonal mode-coupling relation as in Eq. (3.131), the unknown coefficients B_n for $n = I, ..., VI, 1, 2, ...$ are obtained as

$$B_n = \frac{\alpha_2(D_2 p_n^3 - Q_2 p_n)\sinh p_n(H - h)}{ip_n K C_n} + C_n, \qquad (3.134)$$

with $\quad \alpha_2 = 2 \sum_{n=I}^{II} D_2 p_n^3 C_n \sinh p_n(H - h)/\beta_2,$

$$\beta_2 = \sum_{n=I,...,VI,1}^{\infty} \frac{i(D_2 p_n^3 + 2k_{55}ip_n^2)(D_2 p_n^4 - Q_2 p_n)\sinh^2 p_n(H - h)}{K p_n C_n}.$$

Once the constants B_n for $n = I, II$ are determined, the reflection and

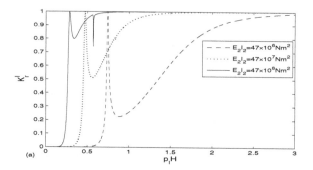

(a)

FIGURE 3.2

Variation of K_r^I versus $p_I H$ with $k_{55} = 0.1E + 11\mathrm{Nm/rad}$ and $h/H = 0.5$ for various values of $E_2 I_2$ with $k_{33} = 0.1E + 4\mathrm{Nm}^{-1}$.

transmission coefficients associated with the wave modes of the floating plate and the articulated plate can be computed using the formulae $K_r^i = (A_i + B_i)/(2C_i)$ and $K_t^i = (A_i - B_i)/(2C_i)$ for $i = I, II$. For an understanding of the wave scattering process, computational results are analysed in two different cases with $E_1 I_1 = 4.7 \times 10^{11}\mathrm{Nm}^2$. In Figure 3.2, the reflection coefficient K_r^I versus $p_I H$ is plotted for different values of $E_2 I_2$ with $k_{55} = 0.1E + 11\mathrm{Nm/rad}$, $k_{33} = 0.1E + 4\mathrm{Nm}^{-1}$ and $h/H = 0.5$. It is observed that for small values of $p_I H$, wave reflection in the floating plate mode is negligible while for large values of $p_I H$, almost full reflection is observed. However, for a moderate wave number, a certain resonating pattern in the reflection coefficient is observed which diminishes as the wave number increases and full reflection is observed for a higher wave number in the floating plate mode. The resonating pattern may be developed due to the interaction of the flexural gravity waves in floating and submerged plate modes for waves of intermediate depth. Further, the resonating pattern in the wave reflection reduces with a decrease in submerged plate rigidity. In Figure 3.3, the reflection coefficient K_r^I versus $p_I H$ is plotted for different values of k_{33} with $k_{55} = 0.1E + 11\mathrm{Nm/rad}$, $E_2 I_2 = 47.0E + 6\mathrm{Nm}^{-1}$ and $h/H = 0.5$. The general pattern is similar to that of Figure 3.2. However, it is observed that with an increase in the spring constant, the reflection coefficient decreases significantly having a resonating pattern for moderate wave numbers. The pattern is similar to wave reflection in the floating elastic plate having a crack in the two-layer fluid as in [10].

3.4.2   Wave scattering by finite flexible horizontal plates

In this section, free surface gravity wave scattering by a floating elastic plate in the presence of a submerged flexible plate is considered. Here, both the plates are assumed to be of finite length and the problems are discussed in

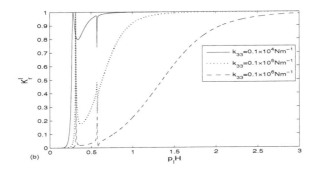

FIGURE 3.3
Variation of K_r^I versus $p_I H$ with $k_{55} = 0.1E + 11\text{Nm/rad}$ and $h/H = 0.5$ for various values of k_{33} with $E_2 I_2 = 47 \times 10^6 \text{Nm}^2$.

the cases of water of both finite and infinite depths. Here, a more general mathematical problem of surface gravity wave interaction with floating elastic plate in the presence of a submerged flexible horizontal plate is formulated under the assumption of the linearised water wave theory in two dimensions. Solution procedure for specific physical problems will be discussed separately. It is assumed that a finite floating flexible plate of finite length occupies the region $-a < x < a, y = 0$ on the mean free surface and a submerged plate is kept in position at $-a < x < a, y = h$ and that $y = H$ is the bottom bed in the case of water of finite depth and $y \to \infty$ in the case of water of infinite depth as in Figure 3.4. Thus, the fluid domains are divided into two regions, namely, the open water region and the floating plate-covered region. It is assumed that ζ_1 and ζ_2 are the deflections of the floating and submerged plates, respectively, and the problems are analysed in the two-dimensional Cartesian coordinate systems. Thus, the spatial velocity potential $\phi(x, y)$ satisfies the Laplace equation as in Eq. (3.99), the bottom boundary conditions are as given by Eqs. (3.100) and (3.101), the boundary conditions in Eqs. (3.102) and (3.103) are satisfied on the floating and submerged plates, and the continuity conditions are at the interface $x = 0$ in Eqs. (3.104) and (3.105). The linearised free surface boundary condition in the open water region is given by

$$\frac{\partial \phi}{\partial y} + K\phi = 0 \quad \text{on } y = 0, \ -\infty < x < -a, \ a < x < \infty. \tag{3.135}$$

The far field radiation conditions for surface wave scattering by the flexible plates are of the form

$$\phi(x, y) = \begin{cases} \{I_0 e^{ik_0 x} + R_0 e^{-ik_0 x}\} Y_0(y) & \text{as} \quad x \to -\infty, \\ T_0 e^{ik_0 x} Y_0(y) & \text{as} \quad x \to \infty, \end{cases} \tag{3.136}$$

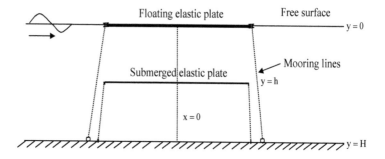

FIGURE 3.4
Schematic diagram for moored floating and submerged flexible plates.

where I_0 is associated with the amplitude of the incident wave and R_0 and T_0 are the amplitudes of the waves associated with the reflected and transmitted waves. In Eq. (3.136), $Y_0(y)$ is given by

$$
Y_0(y) = \begin{cases} \dfrac{\cosh k_0(H - y)}{\cosh k_0 H} & \text{in the case of finite water depth,} \\ e^{-k_0 y} & \text{in the case of infinite water depth,} \end{cases} \tag{3.137}
$$

where k_0 satisfies

$$
K = \begin{cases} k_0 \tanh k_0 H & \text{in the case of finite water depth,} \\ k_0 & \text{in the case of infinite water depth,} \end{cases} \tag{3.138}
$$

with $K = \omega^2/g$. Assuming that both the floating and submerged plates are moored at both ends, the edge conditions at $x = \pm a$ and $y = 0, h$ are given by

$$
\frac{\partial^3 \phi}{\partial x^2 \partial y} = 0 \quad \text{at} \quad (x, y) = (\pm a, d) \tag{3.139}
$$

and

$$
EI \frac{\partial^4 \phi}{\partial x^3 \partial y} = q \frac{\partial^2 \phi}{\partial x \partial y} \quad \text{at} \quad (x, y) = (\pm a, d), \tag{3.140}
$$

where q is the stiffness constant as in [64] and $d = 0, h$. It may be noted that if the stiffness constant $q = 0$, then the floating elastic plate behaves as a plate with a free edge. In order to solve the physical problem at hand, the deflection, slope of deflection, bending moment and shear force acting on the floating and submerged elastic plates are assumed to be continuous at $(0, 0)$ and $(0, h)$ (as in [61]), which yields

$$
\frac{\partial \phi}{\partial y}\bigg|_{(0+,d)} = \frac{\partial \phi}{\partial y}\bigg|_{(0-,d)}, \qquad \frac{\partial^2 \phi}{\partial x \partial y}\bigg|_{(0+,d)} = \frac{\partial^2 \phi}{\partial x \partial y}\bigg|_{(0-,d)}, \tag{3.141}
$$

$$\frac{\partial^3 \phi}{\partial y \partial x^2}\bigg|_{(0+,d)} = \frac{\partial^3 \phi}{\partial y \partial x^2}\bigg|_{(0-,d)}, \quad \frac{\partial^4 \phi}{\partial y \partial x^3}\bigg|_{(0+,d)} = \frac{\partial^4 \phi}{\partial y \partial x^3}\bigg|_{(0-,d)}, \quad (3.142)$$

for $d = 0, h$. Further, the continuity of pressure and velocity at the fluid interface $x = 0, \pm a$, yields

$$\phi \quad \text{and} \quad \phi_x \quad \text{are continuous at} \quad x = 0, \pm a, \; 0 < y < h, h < y < H, \quad (3.143)$$

in water of finite depth and $H \to \infty$ in the case of water of infinite depth. Using the geometrical symmetry of the physical problem, the velocity potential $\phi(x, y)$ is split into the reduced potentials $\varphi(x, y)$ and $\Upsilon(x, y)$ as in Eq. (3.115), which transforms the domain of consideration into problems in quarter plane $(0 < x < \infty, 0 < y < \infty)$ in the case of water of infinite depth and semi-infinite strip $(0 < x < \infty, 0 < y < h)$ in the case of water of finite depth. Further, it may be noted that the reduced potentials $\varphi(x, y)$ and $\Upsilon(x, y)$ satisfy the governing Laplace equation as in Eq. (3.99), the bottom boundary conditions as in Eqs. (3.100)–(3.101), the plate covered boundary condition (3.102) for $0 < x < a$ at $y = 0$, the submerged plate boundary condition as in Eq. (3.103) for $0 < x < a$ at $y = h$, and the boundary condition in the open water regions $-\infty < x < -a, \; a < x < \infty$ as in Eq. (3.135). On the other hand, the continuity of pressure and velocity at the interface $x = 0, \pm a$ in Eq. (3.143) yields

$$\varphi(0, y) = 0 \text{ and } \Upsilon_x(0, y) = 0, \text{ at } x = 0, \pm a. \quad (3.144)$$

The far field conditions in terms of the reduced potentials $\varphi(x, y)$ and $\Upsilon(x, y)$ satisfy

$$\varphi(x, y) \sim \left\{ I_0 e^{-ip_0 x} + A_0 e^{ip_0 x} \right\} \psi_0(y), \quad \text{as} \quad x \to \infty, \quad (3.145)$$

$$\Upsilon(x, y) \sim \left\{ I_0 e^{-ip_0 x} + B_0 e^{ip_0 x} \right\} \psi_0(y), \quad \text{as} \quad x \to \infty, \quad (3.146)$$

where $A_0 = R_0 - T_0$ and $B_0 = R_0 + T_0$. The continuity conditions in Eqs. (3.141) and (3.142) yield

$$\partial_y \varphi(0, d) = 0, \quad \partial^3_{yxx} \varphi(0, d) = 0, \quad (3.147)$$

$$\partial^2_{yx} \Upsilon(0, d) = 0, \quad \partial^4_{yxxx} \Upsilon(0, d) = 0, \quad (3.148)$$

where $d = 0, h$. Further, the edge conditions in Eqs. (3.139) and (3.140) yield

$$\partial^3_{yxx} \varphi(a, c) = 0, \quad E_1 I_1 \partial^4_{yxxx} \varphi(a, c) = q_1 \partial^2_{yx} \varphi(a, c), \quad (3.149)$$

$$\partial^3_{yxx} \Upsilon(a, c) = 0, \quad E_2 I_2 \partial^4_{yxxx} \Upsilon(a, c) = q_2 \partial^2_{yx} \Upsilon(a, c), \quad (3.150)$$

where $c = 0, h$. The reduced velocity potential $\varphi(x, y)$ in water of finite depth is given by

$$\varphi(x, y) = \begin{cases} \displaystyle\sum_{n=I}^{X} C_n \sin p_n x \psi_n(y) + \sum_{n=1}^{\infty} C_n \sin p_n x \psi_n(y), & 0 < x < a, \\[3mm] \displaystyle I_0 e^{-ik_0(x-a)} Y_0(y) + \sum_{n=0,1}^{\infty} A_n e^{ik_n(x-a)} Y_n(y), & a < x < \infty, \end{cases} \quad (3.151)$$

where $\psi_n(y)$ is the same as defined in Eq. (3.113) with p_n satisfying the dispersion relation given in Eq. (3.114). Keeping in mind the realistic nature of the physical problem, it is assumed that the dispersion relation (3.114) has two distinct positive real roots p_n, $n = I, II$, eight complex roots p_n, $n = III, ..., X$ of the form $\pm a \pm ib$ and an infinite number of purely imaginary roots of the form $p_n = i\mu_n$, $n = 1, 2, ...$ (as in [100]). However, in the context of the present study, the contributions from four of the complex roots in the expansion formula are ignored which leads to $C_n = 0$ for $n = VII, ..., X$. On the other hand, in the open water region, the vertical eigenfunctions are given by

$$Y_n(y) = \begin{cases} \dfrac{\cosh k_n(H - y)}{\sqrt{N_n}} & \text{for} \quad n = 0, \\[3mm] \dfrac{\cos k_n(H - y)}{\sqrt{N_n}} & \text{for} \quad n = 1, 2, ..., \end{cases} \qquad (3.152)$$

where $N_n = H/2(1 + \sinh 2k_n H/2k_n H)$ and k_n satisfies the surface gravity wave dispersion relation in k as in Eq. (3.138), which has one real root k_0 and infinitely many imaginary roots of the form ik_n (as discussed in Chapter 1). Further, the vertical eigenfunctions $Y_n(y)$ satisfy the orthogonal mode-coupling relation

$$\langle Y_m, Y_n \rangle = \int_0^H Y_m(y) Y_n(y) dy = \delta_{mn}. \qquad (3.153)$$

Using the continuity of velocity and pressure at $x = a$, (as in Eq. (3.122)) and using the orthogonal mode coupling relation as defined in Eq. (3.153), two systems of equations for the determination of A_n and C_n are obtained as (hereafter, the series for velocity potentials are truncated after N terms)

$$A_m = \sum_{n=0,I,...,VI,1}^{N} C_n \sin p_n a X_{mn} - I_0 \delta_{m0}, \quad m = 0, 1, 2, ..., N, \qquad (3.154)$$

$$ik_m A_m = \sum_{n=0,I,...,VI,1}^{N} p_n C_n \cos p_n a X_{mn} + ik_0 I_0 \delta_{m0}, \quad m = 0, 1, ..., N, \qquad (3.155)$$

with $\quad X_{mn} = \displaystyle\int_0^H \psi_n(y) Y_m(y) dy = \dfrac{-\sinh p_n(H - h) U_{mn} + V_{mn}}{(k_m^2 - p_n^2)\sqrt{N_m}},$

$\qquad U_{mn} = \dfrac{1}{[-\mathcal{D}(ip_n, h)]} \Big[k_m \sinh k_m(H - h)[p_n T_1(ip_n, 2) \sinh p_n h$

$\qquad\qquad + \ K \cosh p_n h] + p_n \cosh k_m(H - h)[-\mathcal{D}(ip_n, h)]$

$\qquad\qquad - \ \{[p_n T_1(ip_n, 2)k_m \sinh k_m H + K p_n \cosh k_m H]\} \Big],$

$\qquad V_{mn} = -\cosh k_m(H - h) \cosh p_m(H - h)$

$\qquad\qquad \times \ [k_m \tanh k_m(H - h) - p_n \tanh p_m(H - h)].$

Thus, the systems of Eqs. (3.154) and (3.155) are solved simultaneously to determine the unknown constants associated with the reduced potential φ. It may be noted that the continuity condition in Eq. (3.147) and the edge condition in Eq. (3.149) will lead to another system of equations from which all the unknown constants can be obtained.

Proceeding in the similar way as in the case of φ, the velocity potential $\Upsilon(x, y)$ is obtained as

$$
\Upsilon(x, y) = \begin{cases} \displaystyle\sum_{n=I}^{IV} G_n \cos p_n x \psi_n(y) + \sum_{n=1}^{\infty} G_n \cos p_n x \psi_n(y), \ 0 < x < a, \\ \displaystyle I_0 e^{-ik_0(x-a)} Y_0(y) + \sum_{n=0,1}^{\infty} B_n e^{ik_n(x-a)} Y_n(y), \ a < x < \infty, \end{cases}
$$

$$(3.156)$$

where p_n and $\psi_n(y)$ are the same as defined in case of φ. Matching the velocity and pressure at $x = a$ and proceeding in a similar manner as in the case of φ, the boundary value problem associated with Υ yields two systems of equations for the determination of B_m, and G_m for $m = 0, 1, ..., N$ are obtained as

$$
B_m = \sum_{n=I,...,IV,1}^{N} G_n \cos p_n a X_{mn} - I_0 \delta_{m0}, \qquad (3.157)
$$

$$
ik_m B_m = \sum_{n=I,...,IV,1}^{N} p_n G_n \sin p_n a X_{mn} + ik_0 I_0 \delta_{m0}, \qquad (3.158)
$$

where X_{mn} is the same as in Eqs. (3.154) and (3.155). It may be noted that the required continuity conditions in Eq. (3.148) and the edge condition in Eq. (3.150) in terms of Υ will lead to another system of equations for complete determination of Υ.

On the other hand, the reduced velocity potentials $\varphi(x, y)$ and $\Upsilon(x, y)$ in the case of infinite water depth are of the forms

$$
\varphi = \begin{cases} \displaystyle\sum_{n=0,I}^{V} C_n \sin k_n x F_n(y) + \int_0^\infty \frac{A(\xi) L(\xi, y) \sinh \xi x d\xi}{\Delta(\xi)}, \quad 0 < x < a, \\ \displaystyle \{I_0 e^{-ik_0(x-a)} + A_0 e^{ik_0(x-a)}\} f_0(y) + \int_0^\infty \frac{M(\xi, y) R(\xi) e^{-\xi x} d\xi}{\Delta(\xi)}, x > a, \end{cases}
$$

$$(3.159)$$

and

$$
\Upsilon = \begin{cases} \displaystyle\sum_{n=0,I}^{V} G_n \cos k_n x F_n(y) + \int_0^\infty \frac{A(\xi) L(\xi, y) \cosh \xi x d\xi}{\Delta(\xi)}, \ 0 < x < a, \\ \displaystyle \{I_0 e^{-ik_0(x-a)} + B_0 e^{ik_0(x-a)}\} f_0(y) + \int_0^\infty \frac{T(\xi) M(\xi, y) e^{-\xi x} d\xi}{\Delta(\xi)}, x > a. \end{cases}
$$

$$(3.160)$$

In Eqs. (3.159) and (3.160), the eigenfunctions $f_0(y)$ and $M(\xi, y)$ are given by

$$f_0(y) = e^{-k_0 y}, M(\xi, y) = \xi \cos \xi y - K \sin \xi y \qquad (3.161)$$

and satisfy the orthogonal relation

$$\int_0^\infty f_0(y) M(\xi, y) dy = 0. \qquad (3.162)$$

On the other hand, the eigenfunctions $L(\xi, y)$ and $F_n(y)$ in (3.159) and (3.160) are the same as defined in Eq. (3.123) of Subsection 3.4.1, with the eigenvalues k_n satisfying the dispersion relation in Eq. (3.114) with $H \to \infty$. The dispersion relation in Eq. (3.114) has eight complex roots of the form $k_n = \pm\alpha \pm i\beta$ for $n = III, IV, ..., X$. As discussed in the case of finite water depth, in the present problem, four complex roots $k_n, n = I, II$, III, IV with positive real parts are considered while the other four complex roots with negative real parts are neglected for the sake of the bounded solution. Using the velocity potentials in each region as defined in Eqs. (3.159) and (3.160) along with the continuity of pressure and velocity at the interface $x = a$ and the edge conditions, the unknown constants and functions will be determined. It may be noted that in the case of infinite water depth, the boundary value problems will lead to a system of integral equations for the determination of the unknown constants and functions associated with the velocity potentials as defined in Eqs. (3.159) and (3.160). The details require special attention as the application of integral equation techniques are involved and are deferred here in the present monograph.

As particular cases of this general problem, surface wave interaction with (i) finite floating flexible plate, (ii) floating flexible membrane, (iii) horizontal submerged flexible plate and (iv) horizontal submerged flexible membrane, with various types of edge conditions, can be obtained in a straightforward manner.

3.5 Examples and exercises

Exercise 3.1 Using Green's function technique as discussed in Subsection 3.2, derive the velocity potentials for the flexural gravity wavemaker problem in cylindrical polar coordinate system in water of finite and infinite depths.

Exercise 3.2 Derive the expansion formulae based on Green's function technique for the flexural gravity wavemaker problem in a two-layer fluid having a flexible plate covered free surface and an interface associated with a two-dimensional Laplace equation and reduced wave equation independently.

Exercise 3.3 Derive the expansion formulae associated with surface wave interaction with a submerged flexible plate based on Green's function technique in water of finite and infinite water depths.

Exercise 3.4 Using the method of splitting as discussed in Section 3.1, derive Green's function for the flexural gravity wavemaker problem in cylindrical and polar co-ordinate system in water of finite and infinite depths.

Exercise 3.5 Using Green's identity, derive the expansion formulae for the velocity potentials associated with the interaction of flexural gravity waves in a two-layer fluid having a plate-covered surface and an interface in water of finite and infinite depths.

Exercise 3.6 Using Green's identity, discuss the procedure of determination of velocity potential associated with surface wave interaction with circular floating plate in water of finite and infinite depths as discussed in Section 3.3.

Exercise 3.7 Using the expansion formulae for wave structure interaction problems, derive the velocity potentials associated with the surface wave interaction with floating/submerged flexible horizontal membranes for attenuation of surface waves near a vertical wall in water of finite and infinite water depths.

Exercise 3.8 Discuss the solution procedure based on the expansion formulae for the interaction of surface waves with a floating flexible circular plate of finite radius in finite and infinite water depths.

Exercise 3.9 Discuss the procedure of response attenuation of a floating flexible circular plate with the help of a submerged plate in finite and infinite water depths.

Exercise 3.10 Using the expansion formulae, discuss the procedure of obtaining the velocity potential associated with the scattering of surface waves with a submerged flexible circular disc in water of finite and infinite depths.

4

Wave interaction with vertical flexible porous structures

4.1 Introduction

There has been a significant interest in the last two decades in the use of vertical flexible structures as breakwaters. These structures provide an alternative to the conventional rigid-fixed breakwaters in areas where poor foundation conditions exist or where protection is required on a temporary basis. These kinds of structures can be quickly deployed, are reusable and can act as low cost wave protection systems compared to those of the rigid-fixed structures. The permeability characteristic of these structures will dramatically reduce the reflected and transmitted wave heights along with the hydrodynamic forces acting on the structures. These kinds of structures can be effectively used for oil spill containment, for temporary protection during coastal construction works, for augmentation of existing breakwaters and for seasonal protection (as in [73], [136] and [152]). Thus, accurate prediction of transmission characteristics and the dynamic response of these flexible porous breakwater are of significant importance in coastal-engineering practices.

There have been numerous investigations, both theoretical and experimental, on flexible breakwaters with and without porosity (see [1], [140], [143], [144], [154] and [155] for details). Apart from wave scattering in a single-fluid medium, there has been certain progress on wave scattering by flexible barriers in a two-layer fluid having free surface and interface (see [73], [136]). Wave past porous screens have significant applications in reducing wave resonance in tanks. The emphasis in this chapter is on briefly discussing the solution procedure for wave interaction with flexible porous vertical structures in single- and double-layer fluids in water of finite and infinite depths which is of recent interest in coastal engineering. The general solution procedures discussed here are based on the application of the eigenfunction expansion method, least square approximations and the wide spacing approximation. The general forms of the expansion of the velocity potentials are mentioned for wave interaction with vertical flexible porous structures in infinite water depth without going into the details of solution. Several problems are left as exercises at the end of the chapter, some of which have not yet been studied in the literature.

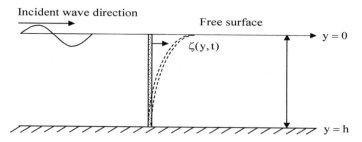

FIGURE 4.1
Schematic diagram of wave scattering by flexible porous barrier.

4.2 Eigenfunction expansion method

In this section, surface gravity wave scattering by vertical flexible porous barriers is discussed in single- and double-layer fluids in finite water depth based on the eigenfunction expansion method. The wave trapping by a vertical flexible porous barrier in finite water depth is analysed. Further, the wave scattering by vertical flexible porous barrier in a two-layer fluid having free surface and interface is analysed. Major emphasis is given to discussing the solution procedures to solve the associated boundary value problems.

4.2.1 Scattering of surface waves by a vertical barrier

In this subsection, the scattering of free surface gravity waves by a vertical flexible porous barrier is analysed in water of finite depth. The structure is assumed to be kept fixed at the bottom and have a free edge at the free surface (as in Figure 4.1). The flexible structure is assumed to be uniform in the longitudinal direction thus allowing two-dimensional analysis. The structural response is assumed to be small to ensure that the small amplitude linear theory is applicable for both the fluid and structure. For analysis, the Cartesian coordinate system with the origin on the mean free surface and the y-axis positive upward is used. The physical problem is considered in the two-dimensional Cartesian coordinate system. Assume that the fluid is inviscid and incompressible and that the motion is irrotational in nature and simple harmonic in time with frequency ω such that the velocity potential is of the form $\Phi(x, y, t) = Re[\phi(x, y)e^{-i\omega t}]$. Thus, the spatial velocity potential $\phi(x, y)$ satisfies

$$\frac{\partial^2 \phi}{\partial x^2} + \frac{\partial^2 \phi}{\partial y^2} = 0 \quad \text{in the fluid domain.} \tag{4.1}$$

The linearised free surface boundary condition in the open water region is given by

$$\frac{\partial \phi}{\partial y} + K\phi = 0 \quad \text{on } y = 0, \ -\infty < x < -\infty. \tag{4.2}$$

The bottom boundary conditions are given by

$$\frac{\partial \phi}{\partial y} = 0 \quad \text{on} \quad y = h \quad \text{in the case of finite depth} \tag{4.3}$$

and

$$\phi, |\nabla \phi| \to 0 \quad \text{as} \quad y \to \infty \quad \text{in the case of infinite depth.} \tag{4.4}$$

The far field radiation conditions for surface wave scattering by the vertical flexible structure are of the form

$$\phi(x, y) = \begin{cases} (I_0 e^{ik_0 x} + R_0 e^{-ik_0 x})\psi_0(y) & \text{as} \quad x \to -\infty, \\ T_0 e^{ik_0 x}\psi_0(y) & \text{as} \quad x \to \infty, \end{cases} \tag{4.5}$$

where I_0 is associated with the amplitude of the incident wave assumed to be known and R_0 and T_0 are the unknown amplitudes of the waves associated with the reflected and transmitted waves. In Eq. (4.5), $\psi_0(y)$ is given by

$$\psi_0(y) = \begin{cases} \dfrac{\cosh k_0(h - y)}{\cosh k_0 h}, & \text{in the case of finite water depth,} \\ e^{-k_0 y} & \text{in the case of infinite water depth,} \end{cases} \tag{4.6}$$

where k_0 satisfies

$$K = \begin{cases} k_0 \tanh k_0 h, & \text{in the case of finite water depth,} \\ k_0 & \text{in the case of infinite water depth,} \end{cases} \tag{4.7}$$

with $K = \omega^2/g$. Assume that the flexible structure is thin and is deflecting horizontally with a displacement $\zeta(y, t) = Re\{\xi(y)e^{-i\omega t}\}$, which is assumed to be small compared to the water depth, with $\xi(y)$ being the maximum amplitude of the deflection of the flexible structure. Further, assume that the flexible structure is porous and $W(y, t) = Re\{w(y)e^{-i\omega t}\}$ is the normal component of the fluid passing through the breakwater. Then the linearised kinematic condition on the mean surface $x = 0$ of the flexible structure yields

$$\frac{\partial \Phi}{\partial x} = W(y, t) + \frac{\partial \zeta}{\partial t} \quad \text{on} \quad x = 0, \ y \in B, \tag{4.8}$$

where $B = (0, h)$ in the case of finite water depth and $B = (0, \infty)$ in the case of infinite water depth. Assuming that the flexible structure has fine pores and that the flow inside the breakwater follows Darcy's law, the porous flow velocity is linearly proportional to the pressure difference between the two sides of the breakwater (as in [22]) which yields

$$W(y, t) = \frac{b}{\mu}(P_1 - P_2) \quad \text{at} \quad x = 0, \tag{4.9}$$

where μ is the dynamic viscosity coefficient, b is a material constant having the dimension of length and P_j is the hydrodynamic pressure exerted on both sides of the wall. From Bernoulli's equation, P_j is given by

$$P_j = -\rho \frac{\partial \Phi_j}{\partial t} \quad (j = 1, 2), \tag{4.10}$$

where ρ is the fluid density which is assumed to be constant. Thus, Eqs. (4.8)–(4.10) yield

$$w(y) = \frac{ib\rho\omega}{\mu}\{\phi_1(0, y) - \phi_2(0, y)\}. \tag{4.11}$$

Thus, Eqs. (4.8)–(4.11) yield the boundary condition on the flexible breakwater as given by

$$\frac{\partial \phi_j}{\partial x} = ik_0 G\{\phi_1(0, y) - \phi_2(0, y)\} - i\omega\xi \quad \text{on} \quad x = 0, \ y \in B, \tag{4.12}$$

where $G = b\rho\omega/\mu k_0$ is known as the porous effect parameter ([18]). More general conditions on the porous structure where G is complex are derived in [153] and are used in the literature (see [120], [152] and [154]).

Next, the structural response of the flexible breakwater is analysed by assuming that the breakwater behaves like an isotropic beam of uniform rigidity EI and uniform mass per unit length m_s subjected to uniform axial force Q (Q being the same as N in Section 1.4.3). Thus, the equation of motion of the flexible vertical structure (which is acted upon by fluid pressure on both sides) is given by (as in [1] and [144])

$$EI\frac{d^4\xi}{dy^4} - Q\frac{d^2\xi}{dy^2} - m_s\omega^2\xi = i\omega\rho[\phi_1(0, y) - \phi_2(0, y)] \quad \text{for} \quad y \in B. \tag{4.13}$$

Assuming that the flexible structure is fixed at seabed and is free at the free surface, the boundary conditions at the two ends of the structure lead to the vanishing of the bending moment and shear force near the free surface and vanishing of the structural deflection and slope of deflection at the bottom bed of the structure which yields

$$\xi(h) = 0, \ \xi_y(h) = 0, \ \xi_{yy}(0) = 0, \ \xi_{yyy}(0) = 0. \tag{4.14}$$

The spatial velocity potential satisfying Eq. (4.1) and the boundary conditions (4.2), (4.3) and (4.5) in the case of water of finite depth is given by

$$\phi(x, y) = \begin{cases} (I_0 e^{ik_0 x} + R_0 e^{-ik_0 x})\psi_0(y) + \sum_{n=1}^{\infty} R_n e^{k_n x}\psi_n(y), & x < 0, \\[4mm] T_0 e^{ik_0 x}\psi_0(y) + \sum_{n=1}^{\infty} T_n e^{-k_n x}\psi_n(y), & x > 0, \end{cases} \tag{4.15}$$

where I_0 is the amplitude of the incident wave assumed to be known and R_n

and T_n are unknown constants to be determined. The vertical eigenfunctions $\psi_n(y)$ are given by

$$\psi_n(y) = \begin{cases} \dfrac{\cosh k_0(h-y)}{\sqrt{N_0}} & \text{for } n = 0, \\ \dfrac{\cos k_n(h-y)}{\sqrt{N_n}} & \text{for } n = 1,2,..., \end{cases} \tag{4.16}$$

where $N_n = h/2(1+\sinh 2k_n h/2k_n h)$ with $k_n = k_0$ for $n = 0$ and $k_n = ik_n$ for $n = 1,2,...$, where k_n satisfies the surface gravity wave dispersion relation in k as in Eq. (3.138) of Chapter 3 (or as in Eq. (1.58) of Chapter 1)). Further, the vertical eigenfunctions $\psi_n(y)$ satisfy the orthogonal relation

$$\langle \psi_m, \psi_n \rangle = \int_0^h \psi_m(y)\psi_n(y)dy = \delta_{mn} \quad \text{for all } m,n = 0,1,..., \tag{4.17}$$

where δ_{mn} is the Kroneckar delta. Using the orthogonal characteristics of $\psi_n(y)$ as in Eq. (4.16) and the continuity of the velocity at $x = 0$, Eq. (4.8) yields

$$R_0 + T_0 = I_0, \quad R_n + T_n = 0, \quad \text{for } n = 1,2,.... \tag{4.18}$$

Substituting for R_n in terms of T_n as in Eq. (4.18), Eqs. (4.15) and (4.13) yield

$$\xi(y) = \sum_{n=1}^{4} C_n f_n(y) + \sum_{n=0}^{\infty} a_n T_n \psi_n(y) + b_0 I_0 \psi_0(y) \tag{4.19}$$

where $f_n(y)$ is given by

$$f_1(y) = \frac{\cosh p_1 y}{\cosh p_1 h}, \ f_2(y) = \frac{\sinh p_2 y}{\sinh p_2 h}, \ f_3(y) = \frac{\cos p_3 y}{\cos p_3 h}, \ f_4(y) = \frac{\sin p_4 y}{\sin p_4 h},$$

with p_n being the roots of the characteristic equation $EIp_n^4 - Qp_n^2 - m_s\omega^2 = 0$ with $p_n = ip_n$ for $n = 3,4$. Further, in Eq. (4.19), a_n and b_0 are given by

$$a_n = \begin{cases} \dfrac{-2i\rho\omega}{EIk_0^4 - Qk_0^2 - m_s\omega^2} & \text{for } n = 0, \\ \dfrac{-2i\rho\omega}{EIk_n^4 + Qk_n^2 - m_s\omega^2} & \text{for } n = 1,2,..., \end{cases} \tag{4.20}$$

$$b_0 = \frac{2i\rho\omega}{EIk_0^4 - Qk_0^2 - m_s\omega^2}. \tag{4.21}$$

However, in Eq. (4.19), C_n and T_n are the unknown constants to be determined. Substituting for $\xi(y)$ from Eq. (4.19) into Eq. (4.12) yields

$$\sum_{n=0}^{\infty} E_n T_n \psi_n(y) + i\omega \sum_{n=1}^{4} \{C_n f_n(y)\} = I_0 h_0 \psi_0(y), \tag{4.22}$$

where E_n and h_0 are given by

$$h_0 = i(2k_0 G - b_0\omega)$$

and

$$E_n = \begin{cases} ik_0(1 + 2G) + i\omega a_0 & \text{for} \quad n = 0, \\ -k_n + 2ik_0 G + i\omega a_n & \text{for} \quad n = 1, 2, \dots. \end{cases} \tag{4.23}$$

Using the orthogonal property of $\psi_n(y)$ as in Eq. (4.17) and truncating the series after N terms, from Eq. (4.22) a system of $N + 1$ equations for T_n is obtained as

$$T_n = \begin{cases} \dfrac{-i\omega}{E_0}\left\{ \displaystyle\sum_{m=1}^{4} C_m \int_0^h f_m(y)\psi_0(y)dy - h_0 I_0 \right\}, & \text{for } n = 0, \\[4mm] \dfrac{-i\omega}{E_n}\left\{ \displaystyle\sum_{m=1}^{4} C_m \int_0^h f_m(y)\psi_n(y)dy \right\}, & \text{for} \quad n = 1, 2, \dots. \end{cases} \tag{4.24}$$

Further, substituting the values of $\xi(y)$ into Eq. (4.14), another system of four equations is obtained. These systems of 4 equations are solved along with the systems of equations in (4.24) to obtain the unknown constants T_n for $n = 0, 1, 2, \dots, N$ and C_n for $n = 1, 2, 3, 4$. Substituting the values of T_n in Eq. (4.18), R_n is obtained and thus the velocity potential $\phi(x, y)$ and plate deflection $\xi(y)$ can be obtained from Eqs. (4.15) and (4.19), respectively. The reflection and transmission coefficients K_r and K_t can be easily computed from the relations $K_r = |R_0/I_0|$ and $K_t = |T_0/I_0|$ and analysed to understand the hydrodynamic characteristics of physical importance. Other important quantities such as the free surface elevation and the hydrodynamic pressure and force acting on the barrier can be obtained from the expression for the velocity potential in Eq. (4.15) (see [1], [115] and [120] for further details on specific physical problems).

4.2.2   Wave trapping by a vertical barrier near a wall

In the case of wave scattering by a vertical flexible porous barrier discussed in the previous section, the barrier is kept at $x = 0$ and the wave field is infinitely extended along the x-axis. On the other hand, in the case of wave trapping, the barrier is kept at a finite distance from a rigid wall located at $x = L$ as in Figure 4.2. As a result, the incident wave of amplitude I_0, after being transmitted by the flexible porous barrier, oscillates between the sea wall and the breakwater which leads to the development of a trapping system. The concept was first experienced by a porous barrier near a wall in [23] and was generalised to flexible porous structures near a wall in [140]. This concept was further generalised to deal with partial barriers near a wall (see [120] and [152]). In this subsection, the trapping of surface waves by a flexible porous barrier near a vertical rigid wall which plays a significant role in coastal

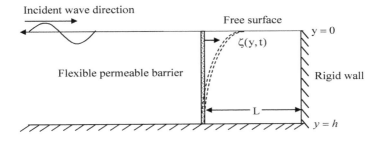

FIGURE 4.2
Schematic diagram of wave trapping by flexible porous barrier.

engineering is analysed. In this case, it is assumed that a monochromatic wave is incident on the flexible porous barrier which is kept near a wall at $x = L$ and the barrier is at $x = 0$. Assuming that the velocity potential is of the type as discussed for the scattering problem, the spatial velocity potential $\phi(x, y)$ satisfies Eq. (4.1) along with the free surface boundary conditions as in Eq. (4.2), the bottom boundary conditions as in Eqs. (4.3) and (4.4) and the boundary conditions as in Eqs. (4.12) and (4.13) along the barrier length B. However, as the flexible barrier is located at a finite distance from a wall, in this case the far field condition is given by

$$\phi(x, y) = \{I_0 e^{ik_0 x} + R_0 e^{-ik_0 x}\}\psi_0(y), \quad \text{as} \quad x \to -\infty, \qquad (4.25)$$

where I_0 is associated with the amplitude of the incident wave assumed to be known and R_0 is the unknown amplitude of the waves associated with the reflected waves. In addition, near the wall there is no flow in the horizontal direction, which yields

$$\frac{\partial \phi}{\partial x} = 0 \quad \text{at} \quad x = L, \text{ for } 0 < y < h. \qquad (4.26)$$

In this case also, under the assumption that the flexible structure is fixed at seabed and is free at the free surface, the boundary conditions at the two ends of the structure yield that ϕ satisfies the boundary condition in Eq. (4.14). The spatial velocity potential satisfying Eq. (4.1) and the boundary conditions in Eqs. (4.2), (4.3), (4.25) and (4.26) are given by

$$\phi(x, y) = \begin{cases} I_0 e^{ik_0 x}\psi_0(y) + \displaystyle\sum_{n=0}^{\infty} R_n e^{-ik_n x}\psi_n(y), & x < 0, \\ \displaystyle\sum_{n=0,1}^{\infty} T_n \cos k_n(L - x)\psi_n(y), & 0 < x < L, \end{cases} \qquad (4.27)$$

where $k_n = k_0$ for $n = 0$, $k_n = ik_n$ for $n = 1, 2, ...$, R_n and T_n are unknown constants to be determined and the vertical eigenfunctions $\psi_n(y)$ are the same as given in Eq. (4.16) satisfying the orthogonal relation as in Eq. (4.17). Using the orthogonal characteristics of $\psi_n(y)$ as in Eq. (4.17), continuity of the velocity at $x = 0$ in Eq. (4.8) yields

$$I_0 - R_0 = -iT_0 \sin k_0 L, \quad R_n + T_n \sinh k_0 L = 0, \quad \text{for} \quad n = 1, 2, ... \quad (4.28)$$

From the first relation in Eq. (4.28), it is clear that $R_0 = I_0$ for

$$L = \frac{n\lambda}{2}, \quad n = 1, 2, ..., \quad (4.29)$$

which ensures that in the presence of a vertical flexible porous barrier, all the incident waves get reflected back when the distance between the vertical wall and the barrier is an integer multiple of half of the wave length. This observation is also true for a nonflexible barrier. It was initially observed in [23] which was later generalised to wave trapping by a flexible porous barrier ([140]). On the other hand, in the case of a rigid porous barrier, from Eq. (4.12), it can be easily derived that (see [120])

$$K_r = \frac{1 - G(1 - i \cot k_0 L)}{1 + G(1 + i \cot k_0 L)}, \quad (4.30)$$

which yields that for $G = 0$, $K_r = 1$. On the other hand, if $L = (2n + 1)\lambda/4$, $(n = 1, 2, ...)$,

$$K_r = \frac{1 - G}{1 + G}. \quad (4.31)$$

Thus, $K_r = 0$ for $G = 1$, which means that when the distance between the channel end wall and the vertical porous barrier is equal to a quarter wave length plus an integer multiple of the half wave length of the incident wave, all the waves are absorbed. Further, for $G \to \infty$, $R_0 = 1$, which means full reflection takes place by the channel end wall for a completely transparent barrier.

Next, the full solution associated with the wave trapping problem is obtained by deriving the velocity potentials completely as in Eq. (4.27). Substituting for R_n in terms of T_n as in Eq. (4.28), Eqs. (4.13) and (4.27) yields

$$\xi(y) = \sum_{n=1}^{4} C_n f_n(y) + \sum_{n=0}^{\infty} a_n T_n \psi_n(y) + b_0 I_0 \psi_0(y), \quad (4.32)$$

where $f_n(y)$ is the same as in Eq. (4.19). Further, in Eq. (4.32), a_n and b_0 are given by

$$a_n = \begin{cases} \dfrac{i\rho\omega v_0}{EIk_0^4 - Qk_0^2 - m_s\omega^2} & \text{for} \quad n = 0, \\[3mm] \dfrac{i\rho\omega v_n}{EIk_n^4 + Qk_n^2 - m_s\omega^2} & \text{for} \quad n = 1, 2, ..., \end{cases} \quad (4.33)$$

$$b_0 = \frac{2i\rho\omega}{EIk_0^4 - Qk_0^2 - m_s\omega^2}, \quad (4.34)$$

where $v_0 = i \sin k_0 L - \cos k_0 L$, $v_n = -\sinh k_n L - \cosh k_n L$. However, in Eq. (4.32), C_n and T_n are the unknown constants to be determined. Substituting for $\xi(y)$ from Eq. (4.32) into Eq. (4.12) yields

$$\sum_{n=0}^{\infty} E_n T_n \psi_n(y) + \sum_{n=1}^{4} C_n f_n(y) = h_0 I_0, \qquad (4.35)$$

where E_n and h_0 are given by

$$E_n = \begin{cases} k_0(\sin k_0 L - iGv_0) + i\omega a_0 & \text{for} \quad n = 0, \\ -k_n \sinh k_n L - ik_0 Gv_n + i\omega a_n & \text{for} \quad n = 1, 2, ..., \end{cases} \qquad (4.36)$$

and

$$h_0 = i(2k_0 G - b_0 \omega),$$

where a_n, v_n and b_0 are the same as in Eqs. (4.33) and (4.34).

Using the orthogonal property of $\psi_n(y)$ as in Eq. (4.17), from Eq. (4.22) and truncating the series after N terms, a system of $N + 1$ equations for T_n is obtained as in Eq. (4.24). Further, substituting the values of $\xi(y)$ from Eq. (4.32) into Eq. (4.14), another system of four equations is obtained. This system of 4 equations is solved along with the system of equations in (4.24) to obtain the unknown constants T_n for $n = 0, 1, 2, ..., N$ and C_n for $n = 1, 2, 3, 4$. Substituting the values of T_n in Eq. (4.28), R_n is obtained and thus the velocity potential $\phi(x, y)$ and plate deflection $\xi(y)$ can be obtained from Eqs. (4.27) and (4.32), respectively. Once the plate deflection and velocity potentials are obtained, other physical quantities such as the wave amplitude in the trapped region, the reflection coefficient, such as wave load on the barrier and rigid walls can be obtained in a straightforward manner.

4.2.3 Wave scattering by double vertical barriers

Another problem of interest is the scattering of surface gravity waves by double flexible porous barriers (as in [1] and [143]). The two barriers are assumed to be of the same materials. It is assumed that the two barriers are located at $x = \pm a$ with $\zeta_j(y, t) = Re\{\xi_j(y)e^{-i\omega t}\}$ for $j = 1, 2$ being the deflection of the flexible barriers and $\xi_j(y)$ being the maximum deflection of the j-th barrier (as in Figure 4.3). Thus, the spatial velocity potential ϕ_j satisfies Eq. (4.1) along with the boundary conditions in Eqs. (4.2), (4.3) and (4.5) in finite water depth. Further, Eqs. (4.12)–(4.13) yield the boundary conditions on the flexible breakwaters given by

$$\frac{\partial \phi_{j+1}}{\partial x} = \frac{\partial \phi_j}{\partial x} = ik_0 G(\phi_j - \phi_{j+1}) - i\omega \xi_j, \text{ at } x = \pm a, j = 1, 2, \qquad (4.37)$$

$$EI \frac{d^4 \xi_j}{dy^4} - Q \frac{d^2 \xi_j}{dy^2} - m_s \omega^2 \xi_j = i\omega \rho(\phi_j - \phi_{j+1}), \text{ at } x = \pm a, j = 1, 2. \qquad (4.38)$$

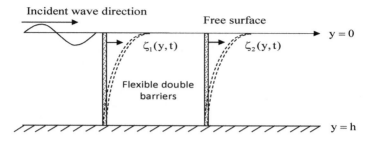

FIGURE 4.3
Schematic diagram of wave scattering by flexible porous barriers.

The far field conditions are given by

$$\phi(x,y) = \begin{cases} (I_0 e^{ik_0(x+a)} + R_0 e^{-ik_0(x+a)})\psi_0(y) & \text{as} \quad x \to -\infty, \\ T_0 e^{ik_0(x-a)}\psi_0(y) & \text{as} \quad x \to \infty. \end{cases} \quad (4.39)$$

Thus, the velocity potentials satisfying Eq. (4.1) along with the boundary conditions (4.2), (4.3) and (4.5) are given by

$$\phi_1(x,y) = I_0 e^{ik_0(x+a)}\psi_0(y) + \sum_{n=0}^{\infty} R_n e^{k_n(x+a)}\psi_n(y), \quad x < -a, \quad (4.40)$$

$$\phi_2(x,y) = \sum_{n=0}^{\infty} a_n(x)\psi_n(y), \quad\quad\quad\quad -a < x < a, \quad (4.41)$$

$$\phi_3(x,y) = \sum_{n=0}^{\infty} T_n e^{-k_n(x-a)}\psi_n(y), \quad\quad\quad\quad x > a, \quad (4.42)$$

where

$$a_n(x) = A_n \frac{\cosh k_n x}{\cosh k_n a} + B_n \frac{\sinh k_n x}{\sinh k_n a}, \quad n = 0, 1, 2, ...,$$

with $k_n = -ik_0$ for $n = 0$. In Eqs. (4.40)–(4.42), R_n, A_n, B_n and T_n for $n = 1, 2, ...$ are unknowns to be determined. Using the continuity of velocity at $x = \pm a$, Eq. (4.37) yields

$$I_0 - R_0 = -i(A_0 \tan k_0 a + B_0 \cot k_0 a), \quad\quad (4.43)$$

$$R_n = -(A_n \tanh k_n a - B_n \coth k_n a), \quad n = 1, 2, ..., \quad (4.44)$$

$$T_0 = i(A_0 \tan k_0 a - B \cot k_0 a), \quad\quad (4.45)$$

$$T_n = -(A_n \tanh k_n a + B_n \coth k_n a), \quad n = 1, 2, ... \quad (4.46)$$

Proceeding in a similar manner as in the case of wave scattering by a single flexible porous barrier, $\xi_j(y)$ is given by

$$\xi_1(y) = \sum_{n=1}^{4} C_n f_n(y) + \sum_{n=0}^{\infty} (\alpha_n A_n + \beta_n B_n)\psi_n(y) + b_0 I_0 \psi_0(y) \quad (4.47)$$

and

$$\xi_2(y) = \sum_{n=1}^{4} D_n f_n(y) + + \sum_{n=0}^{\infty} (\gamma_n A_n + \delta_n B_n) \psi_n(y), \qquad (4.48)$$

where

$$\alpha_n = \begin{cases} \dfrac{i\omega\rho(i\tan k_0 a - 1)}{EIk_0^4 - Qk_0^2 - m_s\omega^2} & \text{for} \quad n = 0, \\[4mm] -\dfrac{i\omega\rho(\tanh k_n a + 1)}{EIk_n^4 + Qk_n^2 - m_s\omega^2} & \text{for} \quad n = 1, 2, ..., \end{cases} \qquad (4.49)$$

$$\beta_n = \begin{cases} \dfrac{i\omega\rho(i\cot k_0 a + 1)}{EIk_0^4 - Qk_0^2 - m_s\omega^2} & \text{for} \quad n = 0, \\[4mm] \dfrac{i\omega\rho(\coth k_n a + 1)}{EIk_n^4 + Qk_n^2 - m_s\omega^2} & \text{for} \quad n = 1, 2, ..., \end{cases} \qquad (4.50)$$

$$b_0 = \dfrac{2i\rho\omega}{EIk_0^4 - Qk_0^2 - m_s\omega^2}, \qquad (4.51)$$

$$\gamma_n = \begin{cases} \dfrac{-i\omega\rho(i\tan k_0 a - 1)}{EIk_0^4 - Qk_0^2 - m_s\omega^2} & \text{for} \quad n = 0, \\[4mm] \dfrac{2i\omega\rho\tanh k_n a}{EIk_n^4 + Qk_n^2 - m_s\omega^2} & \text{for} \quad n = 1, 2, ..., \end{cases} \qquad (4.52)$$

$$\delta_n = \begin{cases} \dfrac{i\omega\rho(i\cot k_0 a + 1)}{EIk_0^4 - Qk_0^2 - m_s\omega^2} & \text{for} \quad n = 0, \\[4mm] \dfrac{-2i\omega\rho\coth k_n a}{EIk_n^4 + Qk_n^2 - m_s\omega^2} & \text{for} \quad n = 1, 2, ... \end{cases} \qquad (4.53)$$

Now, Eq. (4.37) for $j = 1$ yields

$$\frac{\partial \phi_2}{\partial x} = ik_0 G\{\phi_1(x,y) - \phi_2(x,y)\} - i\omega\xi_1, \text{ at } x = -a. \qquad (4.54)$$

Substituting for ϕ_1, ϕ_2 and ξ_1, it can be easily derived that

$$\sum_{n=0}^{\infty} (t_n A_n + s_n B_n) \psi_n(y) + i\omega \sum_{n=1}^{4} C_n f_n(y) + h_0 I_0 \psi_0(y) = 0, \qquad (4.55)$$

where

$$t_n = \begin{cases} k_0 \tan k_0 a + ik_0 G(1 - i\tan k_0 a) + i\omega\alpha_0 & \text{for} \quad n = 0, \\[2mm] -k_n \tanh k_n a + ik_0 G(1 + \tanh k_n a) + i\omega\alpha_n & \text{for} \quad n = 1, 2, ..., \end{cases}$$

$$s_n = \begin{cases} k_0 \cot k_0 a - ik_0 G(1 + i\cot k_0 a) + i\omega\beta_0 & \text{for} \quad n = 0, \\[2mm] k_n \coth k_n a - ik_0 G(1 + \coth k_n a) + i\omega\beta_n & \text{for} \quad n = 1, 2, ..., \end{cases}$$

$$h_0 = i(2k_0 G - b_0\omega).$$

Similarly, Eq. (4.37) for $j = 2$ yields

$$\frac{\partial \phi_3}{\partial x} = ik_0 G\{\phi_2(x,y) - \phi_3(x,y)\} - i\omega\xi_2, \text{ at } x = a. \qquad (4.56)$$

Substituting for ϕ_2, ϕ_3 and ξ_2, it can be easily derived that

$$\sum_{n=0}^{\infty}(u_n A_n + v_n B_n)\psi_n(y) + i\omega\sum_{n=1}^{4}C_n f_n(y) = 0, \qquad (4.57)$$

where $u_n = \begin{cases} -k_0 \tan k_0 a - ik_0 G(1 - i\tan k_0 a) + i\omega\gamma_0, & \text{for} \quad n = 0, \\ k_n \tanh k_n a + 2ik_0 G \tanh k_n a + i\omega\gamma_n, & \text{for} \quad n = 1, 2, ..., \end{cases}$

$v_n = \begin{cases} k_0 \cot k_0 a - ik_0 G(1 + i\cot k_0 a) + i\omega\delta_0, & \text{for} \quad n = 0, \\ k_n \coth k_n a + 2ik_0 G \coth k_n a + i\omega\delta_n, & \text{for} \quad n = 1, 2, ... \end{cases}$

Thus, using the orthogonality of $\psi_n(y)$ from (4.55) and (4.57), it can be easily derived that

$$t_0 A_0 + s_0 B_0 + i\omega\sum_{n=1}^{4}c_n\int_0^h f_n(y)\psi_0(y)dy = -h_0 I_0, \qquad (4.58)$$

$$t_n A_n + s_n B_n + i\omega\sum_{m=1}^{4}c_m\int_0^h f_m(y)\psi_n(y)dy = 0, \qquad (4.59)$$

$$u_0 A_0 + v_0 B_0 + i\omega\sum_{n=1}^{4}c_n\int_0^h f_n(y)\psi_0(y)dy = 0, \qquad (4.60)$$

$$u_n A_n + v_n B_n + i\omega\sum_{m=1}^{4}c_m\int_0^h f_m(y)\psi_n(y)dy = 0. \qquad (4.61)$$

Thus, Eqs. (4.43–4.46) and Eqs. (4.58)–(4.61) yield a system of $4N+4$ equations. Further, using the edge conditions as in Eq. (4.14), another system of 8 equations can be obtained. These sets of equations can be solved for the determination of the unknowns associated with the velocity potentials in Eqs. (4.40)-(4.42) and the plate deflections in Eqs. (4.47) and (4.48).

4.2.4   Wave scattering by vertical barriers in a two-layer fluid

In this subsection, wave scattering by flexible porous vertical barriers in a two-layer fluid having a free surface and interface is briefly discussed in the two-dimensional Cartesian coordinate system. The two fluids are assumed to be inviscid and incompressible and the wave motion is considered in the linearised theory of water wave neglecting the effect of surface tension. The

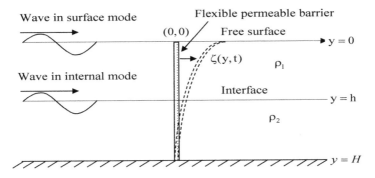

FIGURE 4.4
Schematic diagram of flexible porous barrier in two-layer fluids.

fluid has a free surface (undisturbed surface located at $y = 0$), an interface (undisturbed surface located at $y = h$), each fluid is of infinite horizontal extent occupying the region $-\infty < x < \infty$; $0 < y < h$ in the case of the upper fluid of density ρ_1 and $-\infty < x < +\infty$; $h < y < H$ in the case of the lower fluid of density ρ_2. The undisturbed porous breakwater is placed at $x = 0$; $0 \le y \le H$ as in Figure 4.4. The flow is assumed to be irrotational and simple harmonic in time with angular frequency ω. Thus, like wave scattering by a vertical flexible porous barrier in single homogeneous fluid medium, in this case the velocity potentials Φ_j exist such that $\Phi_j(x, y, t) = Re\{\phi_j(x, y)e^{-i\omega t}\}$, where the spatial velocity potentials $\phi_j(x, y)$, $(j = 1, 2)$ satisfy the Laplace equation

$$\nabla^2 \phi_j = 0 \text{ in the fluid region } j, \qquad (4.62)$$

where $j = 1$ refers to the fluid region $-\infty < x < 0$; $0 < y < H$ and $j = 2$ refers to the fluid region $0 < x < +\infty$; $0 < y < H$ in water of finite depth and accordingly $H \to \infty$ in water of infinite depth. Further, ϕ_j satisfies the linearised free surface condition in Eq. (4.2). At the interface, the continuity of the vertical component of velocity and pressure yields the boundary conditions (as discussed in Chapter 1)

$$\left(\frac{\partial \phi_j}{\partial y}\right)_{y=h_+} = \left(\frac{\partial \phi_j}{\partial y}\right)_{y=h_-} \qquad (4.63)$$

and $\left(\dfrac{\partial \phi_j}{\partial y} + K\phi_j\right)_{y=h_+} = s\left(\dfrac{\partial \phi_j}{\partial y} + K\phi_j\right)_{y=h_-}, \ (j = 1, 2),$ $\qquad (4.64)$

where $s = \rho_1/\rho_2$ with $0 < s < 1$. The condition on the rigid bottom is given by

$$\frac{\partial \phi_j}{\partial y} = 0 \ (j = 1, 2) \text{ on } y = H \text{ in water of finite depth.} \qquad (4.65)$$

On the other hand, in water of infinite depth, ϕ satisfies Eq. (4.4). The radiation conditions are given by

$$\phi_1 \to \sum_{n=I}^{II} (I_n e^{ik_n x} + R_n e^{-ik_n x})\psi_n(y) \quad \text{as } x \to -\infty, \qquad (4.66)$$

and

$$\phi_2 \to \sum_{n=I}^{II} T_n e^{ik_n x} \psi_n(y) \quad \text{as } x \to +\infty, \qquad (4.67)$$

where I_n, R_n and T_n for $n = I, II$ are the incident, reflected and transmitted wave amplitudes in surface mode (SM) and internal mode (IM), respectively. It may be noted that k_I and k_{II} are wave numbers for the incident waves in SM and IM, respectively, as in [73]. The expression for ψ_n is given by

$$\psi_n(y) = \begin{cases} \dfrac{g}{\omega}(k_n \cosh k_n y - K \sinh k_n y), & 0 < y < h, \\ \dfrac{\omega}{\sinh k_n(H-h)} \cosh k_n(H-y), & h < y < H. \end{cases} \qquad (4.68)$$

Assuming that the breakwater is deflected horizontally with displacement $\zeta(y,t) = Re[\xi(y)e^{-i\omega t}]$ and $\xi(y)$ denotes the complex deflection amplitude, the boundary condition on the flexible porous breakwater is given by

$$\frac{\partial \phi_j}{\partial x} = ik_I G(\phi_2 - \phi_1) - i\omega\xi \ (j = 1, 2) \text{ on } x = 0; \ 0 < y < H, \qquad (4.69)$$

where $G = b\rho\omega/\mu k_I$ is a complex porous-effect parameter in the two-layer fluid (as in [153]). The breakwater response is analysed by assuming that the breakwater behaves like a one-dimensional beam of uniform flexural rigidity EI and mass per unit length m_s. Hence, the governing equation of the breakwater response is given by Eq. (4.13). Assuming that the barrier behaves like a cantilever, as it is assumed in the study, and that the breakwater has free and fixed ends at the free surface and seabed, respectively, the corresponding end conditions are given by

$$\xi''(0) = 0, \ \xi'''(0) = 0, \ \xi(H) = 0, \ \xi'(H) = 0. \qquad (4.70)$$

The continuity deflection, slope of deflection, bending moment and shear force acting on the breakwater at the interface (as in [152]), yields

$$\xi(y), \ \xi'(y), \ \xi''(y), \ \xi'''(y) \text{ which are continuous at } y = h. \qquad (4.71)$$

The spatial velocity potentials ϕ_j for $j = 1, 2$ satisfying Eq. (4.62) along with conditions (4.2) and (4.63)–(4.67) in finite water depth is expressed as

$$\phi_1 = \sum_{n=I}^{II} I_n e^{ik_n x} \psi_n(y) + \sum_{n=I,II,1}^{\infty} R_n e^{-ik_n x} \psi_n(y), \text{ for } x < 0, \qquad (4.72)$$

$$\phi_2 = \sum_{n=I,II,1}^{\infty} T_n e^{ik_n x} \psi_n(y), \qquad \qquad \text{for } x > 0, \qquad (4.73)$$

where the expression for $\psi_n(y)$ for $n = I, II, 1, 2, ...$ is the same as defined in Eq. (4.68). The unknowns R_n and T_n for $n = I, II, 1, 2, ...$ are constants to be determined and k_n for $n = I, II, 1, 2, 3, ...$ is the root of the dispersion relation in p as given by

$$(1-s)p^2 \tanh p(H-h) \tanh ph - pK\{\tanh ph + \tanh p(H-h)\}$$
$$+ K^2\{s \tanh p(H-h) \tanh ph + 1\} = 0. \quad (4.74)$$

The eigenfunctions $\psi_n(y)$ are integrable in $0 < y < H$ having a single discontinuity at $y = h$ and are orthogonal with respect to the inner product as defined by

$$< \psi_m, \psi_n > = \rho_1 \int_0^h \psi_m \psi_n dy + \rho_2 \int_h^H \psi_m \psi_n dy, \quad m, n = I, II, 1, 2, ... \quad (4.75)$$

Further, it may be noted that the orthogonal relation in Eq. (4.75) reduces to the usual one of the single-layer fluid when $\rho_1 = \rho_2$ (details on orthogonality on two-layer fluids are discussed in Section 2.6. and see [73], [88] and [136] for further details). Applying the continuity of ϕ_x in Eq. (4.69) along the porous breakwater at $x = 0$ and invoking the orthogonality of the eigenfunctions as in Eq. (4.75) over $0 < y < H$, it can be easily derived that

$$I_n - R_n = T_n \quad \text{for} \quad n = I, II, \text{ and } R_n = -T_n \quad \text{for} \quad n = 1, 2, 3, ... \quad (4.76)$$

A general solution for the breakwater governing equation is of the form

$$\xi(y) = \sum_{n=1}^{4} C_n f_n(y) + 2 \sum_{n=I}^{II} I_n a_n \psi_n(y) - \sum_{n=I,II,1}^{\infty} a_n T_n \psi_n(y), \ 0 < y < H,$$
$$(4.77)$$

where a_n, C_n and ρ are given by

$$a_n = \begin{cases} \dfrac{i\rho\omega}{EIk_n^4 - Qk_n^2 - m_s\omega^2}, & \text{for } n = I, II, \\[3mm] \dfrac{i\rho\omega}{EIk_n^4 + Qk_n^2 - m_s\omega^2}, & \text{for } n = 1, 2, ..., \end{cases} \quad (4.78)$$

$$C_n = \begin{cases} A_n, \text{ for } 0 < y < h, \\ B_n, \text{ for } h < y < H, \end{cases} \quad \rho = \begin{cases} \rho_1, \text{ for } 0 < y < h, \\ \rho_2, \text{ for } h < y < H, \end{cases} \quad (4.79)$$

with f_n being the same as in Eq. (4.19). In Eq. (4.77), C_n and R_n are to be determined. Substituting for $\xi(y)$ from Eq. (4.77) into Eq. (4.69) yields

$$\sum_{n=I,II,1}^{\infty} E_n T_n \psi_n(y) + i\omega \sum_{n=1}^{4} C_n f_n(y) = \sum_{n=I}^{II} I_n h_n \psi_n(y), \quad (4.80)$$

where E_n and h_n are given by

$$E_n = \begin{cases} i(k_n - 2k_I G) - 2i\omega a_n & \text{for} \quad n = I, II, \\ -k_n - 2ik_0 G - 2i\omega a_n & \text{for} \quad n = 1, 2, ... \end{cases} \quad (4.81)$$

and
$$h_n = -2i(k_I G + \omega a_n).$$

Using the orthogonal property of $\psi_n(y)$ as in Eq. (4.75), from Eq. (4.80) and truncating the series after N terms, a system of $N + 2$ equations for T_n is obtained as

$$T_n = \begin{cases} \dfrac{-i\omega}{E_n} \left\{ \displaystyle\sum_{m=1}^{4} C_m \int_0^h f_m(y)\psi_n(y)dy - h_n I_n \right\}, & \text{for } n = I, II, \\[4mm] \dfrac{-i\omega}{E_n} \left\{ \displaystyle\sum_{m=1}^{4} C_m \int_0^h f_m(y)\psi_n(y)dy \right\}, & \text{for } n = 1, 2, ... \end{cases} \quad (4.82)$$

Further, substituting the values of $\xi(y)$ from Eq. (4.77) into (4.70) and (4.71), another system of eight equations is obtained. This system of 8 equations is solved along with the system of equations in (4.82) to obtain the unknown constants T_n for $n = I, II, 1, 2, ..., N$ and C_n for $n = 1, 2, 3, 4$. Substituting the values of T_n into Eq. (4.76), R_n is obtained, and thus the velocity potential $\phi(x, y)$ and plate deflection $\xi(y)$ can be obtained from Eqs. (4.72), (4.73) and (4.77), respectively. The reflection and transmission coefficients K_r and K_t can be easily computed from the relations $K_r^n = |R_n/I_n|$ and $K_t^n = |T_n/I_n|$ for $n = I, II$ and analysed to understand the hydrodynamic characteristics of physical importance. Other important quantities such as the free surface and interface elevations and the hydrodynamic pressure and force acting on the barrier can be obtained from the expression for the velocity potential in Eqs. (4.72) and (4.73) (see [1], [115], [120] for further details on specific physical problems).

4.3   Method of least square approximation

In the previous section, the eigenfunction expansion method has been applied to deal with the surface gravity wave scattering by a flexible porous vertical barrier where the barrier is extended from the free surface to bottom, which is often referred as a complete barrier. On the other hand, in the case of barriers which are surface piercing partially immersed, bottom standing partial barriers, or barriers which are fully submerged, the orthogonal characteristic of the vertical eigenfunction cannot be applied directly over the length of the barrier due to the discontinuity of the velocity and the existence of singularity at the submerged edge of the barrier. One of the easiest methods to deal with such problems is the least square approximation method, in which the singular behavior at the edge is taken care of due to the integrable nature of the singularity of the velocity at the submerged edge (see [25], [120]). Here, the

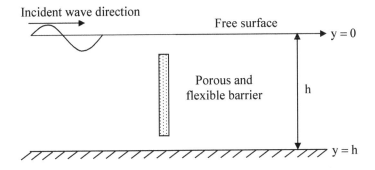

FIGURE 4.5
Schematic diagram of flexible porous partial barrier.

least square approximation method is illustrated through certain problems of
wave scattering by partial flexible porous barriers.

4.3.1 Wave scattering by partial barrier

The study of surface wave scattering by partial barriers is well known in
coastal engineering problems. In the case of deep water waves of large depth
where wave energy concentration is significant near the free surface, surface
piercing partial barriers play a significant role. On the other hand, in case
of long waves where wave energy propagation is uniform in the horizontal
direction, the bottom standing barrier is more important. In this subsection,
surface wave scattering by partial flexible porous barrier is analysed. In this
case it is assumed that a submerged barrier occupies the region $B \equiv \{a <
y < b, x = 0\}$ along the length of the barrier and the region $G \equiv \{0 < y <
a, b < y < h, x = 0\}$ along the gap (as in Figure 4.5). Thus, the boundary value
problem in terms of the spatial velocity potential $\phi(x, y)$ satisfies the governing
equation as in Eq. (4.1) along with the free surface boundary condition as in
Eq. (4.2), the bottom boundary conditions as in Eqs. (4.3) and (4.4) and
the radiation condition as in Eq. (4.5). The velocity potential satisfies the
boundary conditions as in Eqs. (4.12) and (4.13) along the barrier length B.
However, along the gap G, the continuity of pressure and velocity yields

$$\phi, \phi_x \quad \text{are continous along the gap } G. \tag{4.83}$$

At the submerged edge of the barrier,

$$r^{1/2}\nabla\phi \quad \text{is bounded as} \quad r = \{x^2 + (y - c)^2\}^{1/2} \to 0, \tag{4.84}$$

where $(x, y) = (0, c)$ is assumed to be the submerged edge of the barrier.
Further, at the edges, appropriate edge conditions which are similar to the

conditions in Eq. (4.14) have to be prescribed to keep the barrier in position (some of the important edge conditions are briefly discussed in Section 1.4 of Chapter 1).

In the case of wave scattering by partial flexible barriers in water of finite depth, the form of the velocity potential is the same as in Eq. (4.15). Using the continuity of velocity potential along the barrier and the gap, it can be easily derived that R_n and T_n satisfy the relations in Eq. (4.18). Substituting for R_n in terms of T_n, the expression for $\xi(y)$ can be obtained as in Eq. (4.32). Thus, using the continuity of velocity potential along the gap as in Eq. (4.83) and the condition on the flexible porous barrier as in Eq. (4.12) along with $\xi(y)$ as in Eq. (4.13), it can be easily derived that

$$\sum_{n=0}^{\infty} T_n g_n(y) + h_0(y) = 0, \quad 0 < y < h, \tag{4.85}$$

where

$$g_0(y) = \begin{cases} -\psi_0(y), & \text{for} \quad y \in G, \\ E_0 \psi_0(y), & \text{for} \quad y \in B, \end{cases} \tag{4.86}$$

$$g_n(y) = \begin{cases} -\psi_n(y), & \text{for} \quad y \in G, \\ E_n \psi_n(y), & \text{for} \quad y \in B, \end{cases} \tag{4.87}$$

$$h_0(y) = \begin{cases} I_0 \psi_0(y), & \text{for} \quad y \in G, \\ i\omega \sum_{n=1}^{4} C_n f_n(y) - I_0 h_0 \psi_0(y), & \text{for} \quad y \in B, \end{cases} \tag{4.88}$$

where E_n is the same as in Eq. (4.36). Setting

$$S_N(y) = \sum_{n=0}^{N} T_n g_n(y) + h_0(y), \quad 0 < y < h, \tag{4.89}$$

the least square approximation method yields

$$\int_0^h |S_N(y)|^2 dy = \text{minimum.} \tag{4.90}$$

Minimising the integral in Eq. (4.90) with respect to T_n gives

$$\int_0^h S_N^*(y) S_{T_n}(y) dy = 0, \quad (n = 0, 1, 2, ..., N), \tag{4.91}$$

where $*$ denotes the complex conjugate and $S_{T_n}(y)$ is the derivative of $S_N(y)$ with respect to T_n. From Eq. (4.91), a system of equations for determination of T_n is obtained as

$$\sum_{m=0}^{N} T_m^* X_{mn} = b_n, \quad (n = 0, 1, 2, ..., N), \tag{4.92}$$

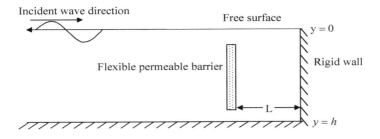

FIGURE 4.6
Schematic diagram of vertical flexible porous barrier.

with

$$b_n = - \int_0^h h_0^*(y) g_n(y) dy, \quad (n = 0, 1, 2, ..., N), \tag{4.93}$$

$$X_{mn} = \int_0^h g_m^*(y) g_n(y) dy, \quad (m, n = 0, 1, 2, ..., N). \tag{4.94}$$

Further, appropriate edge conditions (as discussed in Subsection 1.4.2 of Chapter 1) similar to Eq. (4.14) at the edges of the vertical structures will be used to obtain another set of equations. This set of equations along with the system of equations in Eq. (4.92) will be solved together to determine all the unknowns T_n and C_n from which R_n can be obtained using Eq. (4.18). Substituting for R_n, T_n and C_n in Eqs. (4.15) and (4.19), the velocity potential and plate deflection can be obtained from which other important physical quantities can be computed in a straightforward manner.

4.3.2 Wave trapping by partial barrier near a wall

The phenomenon of wave trapping by a partial flexible porous barrier near a vertical wall is similar to that of a flexible porous barrier which is extended from the free surface to the bottom as in Figure 4.6. Thus, the boundary value problem in terms of the spatial velocity potential $\phi(x, y)$ satisfies the governing equation as in Eq. (4.1) along with the free surface boundary condition as in Eq. (4.2), the bottom boundary conditions as in Eqs. (4.3) and (4.4), the radiation condition as in Eq. (4.25) and the wall boundary condition as in Eq. (4.26). The velocity potential satisfies the boundary conditions as in Eqs. (4.12) and (4.13) along the barrier length B. However, along the gap G, the continuity of pressure and velocity yield that ϕ satisfies Eq. (4.80) and the edge condition as in Eq. (4.81). Further, at the edges, appropriate edge conditions have to be prescribed to keep the barrier in position which is similar to the conditions prescribed in Eq. (4.14). Thus, in the case of wave trapping by a partial flexible barrier in water of finite depth, the velocity potential is of the form given in Eq. (4.27). Further, the continuity of velocity along the barrier and the gap (as in Eq. (4.12)) yield R_n and T_n satisfy the relation in Eq. (4.28), which will lead to the observation that $R_0 = I_0$ for $L = n\lambda/2$ as in the case of a complete barrier (as in Eq. (4.29)). Further, in this case also, near the submerged edge, the spatial velocity potential $\phi(x, y)$ satisfies Eq. (4.14). Substituting for R_n in terms of T_n, the expression for $\xi(y)$ can be obtained as in Eq. (4.32). Thus, using the continuity of velocity potential along the gap as in Eq. (4.83) and the condition on the flexible porous barrier as in Eq. (4.12), along with $\xi(y)$ as in Eq. (4.32), it can be easily derived that

$$\sum_{n=0}^{\infty} T_n g_n(y) + h_0(y) = 0, \quad 0 < y < h, \tag{4.95}$$

where

$$g_0(y) = \begin{cases} v_0 \psi_0(y), & \text{for } y \in G, \\ E_0 \psi_0(y), & \text{for } y \in B, \end{cases} \tag{4.96}$$

$$g_n(y) = \begin{cases} v_n \psi_n(y), & \text{for } y \in G, \\ E_n \psi_n(y), & \text{for } y \in B, \end{cases} \tag{4.97}$$

$$h_0(y) = \begin{cases} 2I_0 \psi_0(y), & \text{for } y \in G, \\ i\omega \sum_{n=1}^{4} C_n f_n(y) - I_0 h_0 \psi_0(y), & \text{for } y \in B, \end{cases} \tag{4.98}$$

with E_n being the same as in Eq. (4.36). Setting

$$S_N(y) = \sum_{n=0}^{N} T_n \psi_n(y) + h_0(y), \quad 0 < y < h, \tag{4.99}$$

the least square approximation method yields a system of equations for determination of T_n as

$$\sum_{m=0}^{N} T_m^* X_{mn} = b_n, \quad (n = 0, 1, 2, ..., N), \tag{4.100}$$

with

$$b_n = \int_0^h h_0^*(y) g_n(y) dy, \quad (n = 0, 1, 2, ..., N), \tag{4.101}$$

$$X_{mn} = \int_0^h g_m^*(y) g_n(y) dy, \quad (m, n = 0, 1, 2, ..., N). \tag{4.102}$$

Further, suitable edge conditions similar to Eq. (4.14) at the edges of the vertical structures will be used to obtain another set of equations. This set of equations along with the system of equations in Eq. (4.100) will be solved together to determine all the unknowns T_n, C_n and D_n together which in turn will determine the velocity potential and plate deflection completely.

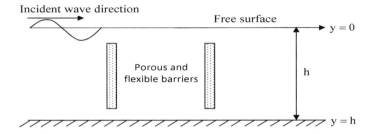

FIGURE 4.7
Schematic diagram of wave scattering by vertical flexible porous barriers.

4.3.3 Wave scattering by two partial barriers

Another efficient method for creating a tranquility zone for various marine activities is the use of double flexible porous barriers. Bottom standing barriers are more suitable at locations where the bottom soil condition is very poor. On the other hand, in the case of large water depth and locations where wave energy concentration is more at the free surface, surface piercing barriers are more suitable. In the present subsection, surface wave scattering by two flexible porous barriers located at $x = \pm a$ will be discussed as in Figure 4.7 like the case of double barriers which extend from free surface to the bottom. In this case, the velocity potentials ϕ_j satisfy Eq. (4.1) along with the free surface boundary condition as in Eq. (4.2) and the bottom boundary conditions in Eqs. (4.3) and (4.5) in the case of finite water depth. The boundary conditions along the flexible barriers are the same as in Eqs. (4.37) and (4.38). Further, along the gap, the continuity of velocity and pressure yields

$$\phi_j, \ \phi_{jx} \quad \text{are continuous along the gap } G. \tag{4.103}$$

At the submerged edges of the barriers,

$$r^{1/2}\nabla\phi_j \quad \text{is bounded as} \quad r = \{(x \pm a)^2 + (y - c)^2\}^{1/2} \to 0, \tag{4.104}$$

where $(0, c)$ is the edge of the barrier. In this case, the velocity potentials ϕ_j are of the form as in Eqs. (4.40)–(4.42). Further, using the continuity of flow velocity along the barrier and the gap yields that R_0, T_0, R_n, T_n, A_n and B_n satisfy the Eqs. (4.43)–(4.46). Further, the plate response $\xi_j(y)$ is of the form as in Eqs. (4.47)–(4.48). Finally, using the boundary conditions along the barriers in Eqs. (4.37) and (4.38) and along the gap as in Eq. (4.103), it can be easily derived that

$$\sum_{n=0}^{\infty}\{A_n p_n(y) + B_n q_n(y)\} + h_0(y) = 0, \ y \in B \cup G \tag{4.105}$$

and

$$\sum_{n=0}^{\infty}\{A_n\tilde{p}_n(y) + B_n\tilde{q}_n(y)\} + \tilde{h}_0(y) = 0, \ y \in B \cup G, \tag{4.106}$$

where

$$p_0(y) = \begin{cases} (i\tan k_0 a - 1)\psi_0(y), & \text{for} \quad y \in G, \\ t_0\psi_0(y), & \text{for} \quad y \in B, \end{cases} \tag{4.107}$$

$$p_n(y) = \begin{cases} -(\tanh k_n a + 1)\psi_n(y), & \text{for} \quad y \in G, \\ t_n\psi_n(y), & \text{for} \quad y \in B, \end{cases} \tag{4.108}$$

$$q_0(y) = \begin{cases} (i\cot k_0 a + 1)\psi_0(y), & \text{for} \quad y \in G, \\ s_0\psi_0(y), & \text{for} \quad y \in B, \end{cases} \tag{4.109}$$

$$q_n(y) = \begin{cases} (\coth k_n a + 1)\psi_n(y), & \text{for} \quad y \in G, \\ s_n\psi_n(y), & \text{for} \quad y \in B, \end{cases} \tag{4.110}$$

$$h_0(y) = \begin{cases} 2I_0\psi_0(y), & \text{for} \quad y \in G, \\ i\omega\sum_{n=1}^{4} C_n f_n(y) + I_0 h_0\psi_0(y), & \text{for} \quad y \in B, \end{cases} \tag{4.111}$$

$$\tilde{p}_0(y) = \begin{cases} -(i\tan k_0 a - 1)\psi_0(y), & \text{for} \quad y \in G, \\ u_0\psi_0(y), & \text{for} \quad y \in B, \end{cases} \tag{4.112}$$

$$\tilde{p}_n(y) = \begin{cases} -2\tanh k_n a\psi_n(y), & \text{for} \quad y \in G, \\ u_n\psi_n(y), & \text{for} \quad y \in B, \end{cases} \tag{4.113}$$

$$\tilde{q}_0(y) = \begin{cases} (i\cot k_0 a + 1)\psi_0(y), & \text{for} \quad y \in G, \\ v_0\psi_0(y), & \text{for} \quad y \in B, \end{cases} \tag{4.114}$$

$$\tilde{q}_n(y) = \begin{cases} -2\coth k_n a\psi_n(y), & \text{for} \quad y \in G, \\ v_n\psi_n(y), & \text{for} \quad y \in B, \end{cases} \tag{4.115}$$

$$\text{and} \quad \tilde{h}_0(y) = \begin{cases} 0, & \text{for} \quad y \in G, \\ i\omega\sum_{n=1}^{4} D_n f_n(y), & \text{for} \quad y \in B. \end{cases} \tag{4.116}$$

Proceeding in a similar manner as in the case of wave scattering by a single partial flexible barrier, applying least square approximation, two sets of equations for determination of A_n, B_n for $n = 1, 2, ...$ are obtained as

$$\sum_{m=0}^{N}(A_m^* X_{mn} + B_m^* Y_{mn}) = b_n, \quad (n = 0, 1, 2, ..., N), \tag{4.117}$$

with

$$b_n = -\int_0^h h_0^*(y)p_n(y)dy, \quad (n = 0, 1, 2, ..., N), \quad (4.118)$$

$$X_{mn} = \int_0^h p_m^*(y)p_n(y)dy, \quad (m, n = 0, 1, 2, ..., N), \quad (4.119)$$

$$Y_{mn} = \int_0^h q_m^*(y)p_n(y)dy, \quad (m, n = 0, 1, 2, ..., N), \quad (4.120)$$

and

$$\sum_{m=0}^N (A_m^* \tilde{X}_{mn} + B_m^* \tilde{Y}_{mn}) = \tilde{b}_n, \quad (n = 0, 1, 2, ..., N), \quad (4.121)$$

with

$$\tilde{b}_n = -\int_0^h \tilde{h}_0^*(y)\tilde{p}_n(y)dy, \quad (n = 0, 1, 2, ..., N), \quad (4.122)$$

$$\tilde{X}_{mn} = \int_0^h \tilde{p}_m^*(y)\tilde{p}_n(y)dy, \quad (m, n = 0, 1, 2, ..., N). \quad (4.123)$$

$$\tilde{Y}_{mn} = \int_0^h \tilde{q}_m^*(y)\tilde{p}_n(y)dy, \quad (m, n = 0, 1, 2, ..., N). \quad (4.124)$$

Further, appropriate edge conditions at the two ends of the barriers similar to the edge conditions assumed in Eq. (4.14) will yield a system of 8 equations, which are solved along with the two systems of equations in Eqs. (4.117) and (4.121) to determine the set of unknowns A_n, B_n for $n = 0, 1, 2, ..., N$ and C_n, D_n for $n = 1, 2, 3, 4$.

4.4 Transform method

In the previous two sections, the wave structure interaction problems are illustrated in the Cartesian coordinate system in the case of finite water depth which has enabled the use of the eigenfunction expansion method. In this section, the form of the velocity potentials dealing with surface gravity wave scattering by vertical flexible porous barrier structures in water of infinite depth is briefly discussed. In this case, the final solution will reduce to a set of integral equations for the determination of the unknowns and details are deferred here. The velocity potentials for the wave scattering by a single flexible porous barrier (in the cases of both complete and partial barriers) in infinite water depth are of the form (as discussed in Chapter 2)

$$
\phi =
\begin{cases}
(I_0 e^{ik_0 x} + R_0 e^{-ik_0 x})\psi_0(y) + \displaystyle\int_0^\infty \frac{A(\xi)M(\xi,y)e^{\xi x}d\xi}{\xi^2 + K^2}, & x < 0, \\[4mm]
T_0 e^{i\mu x}\psi_0(y) + \displaystyle\int_0^\infty \frac{M(\xi,y)B(\xi)e^{-\xi x}d\xi}{\xi^2 + K^2}, & x > 0.
\end{cases}
\tag{4.125}
$$

In Eq. (4.125), the eigenfunctions $\psi_0(y)$ and $M(\xi,y)$ given by

$$
\psi_0(y) = e^{-k_0 y}, \quad M(\xi,y) = \xi \cos \xi y - K \sin \xi y
\tag{4.126}
$$

satisfy the orthogonal relation

$$
\int_0^\infty \psi_0(y)M(\xi,y)dy = 0.
\tag{4.127}
$$

Using the orthogonal characteristics of $\psi_0(y)$ and $M(\xi,y)$ in Eq. (4.127) and the continuity of the velocity at $x = 0$ in Eq. (4.12), from Eq. (4.125) it can be easily derived that

$$
R_0 + T_0 = I_0, \quad A(\xi) + B(\xi) = 0, \quad \text{for all} \quad \xi.
\tag{4.128}
$$

On the other hand, the velocity potential associated with the trapping of surface gravity wave by a flexible porous barrier (in the cases of both complete and partial barriers) near a wall is of the form given by

$$
\phi =
\begin{cases}
\{I_0 e^{ik_0 x} + R_0 e^{-ik_0 x}\}\psi_0(y) + \displaystyle\int_0^\infty \frac{A(\xi)M(\xi,y)e^{-\xi x}d\xi}{\xi^2 + K^2}, & x < 0, \\[4mm]
T_0 \cos k_0(L - x)\psi_0(y) + \displaystyle\int_0^\infty \frac{B(\xi)M(\xi,y)\cosh \xi(L-x)d\xi}{\xi^2 + K^2}, & 0 < x < L.
\end{cases}
\tag{4.129}
$$

In this case also, using the orthogonal characteristics of $\psi_0(y)$ and $M(\xi,y)$ as in Eq. (4.127) and the continuity of the velocity at $x = 0$ in Eq. (4.12), Eq. (4.125) yields

$$
I_0 - R_0 = -iT_0 \sin k_0 L, \quad A(\xi) + B(\xi) \sinh \xi L = 0, \quad \text{for all} \quad \xi.
\tag{4.130}
$$

From the first relation of Eq. (4.130), it is evident that full reflection will take place in the case of infinite water depth when the distance between the vertical wall and the barrier is an integer multiple of half of the wave length. This observation is similar to the observation in the case of finite water depth. Further, from Eqs. (4.12) and (4.125), it can be easily derived that for a porous rigid complete barrier, Eq. (4.30) is satisfied in the case of infinite water depth which ensures all the incident waves will be absorbed and no wave reflection will take place in water of infinite depth. Similar conditions are discussed in water of finite depth. Using the orthogonal characteristic of the eigenfunctions

$\psi_0(y)$ and $M(\xi, y)$ as in Eq. (4.127), it can be easily derived that Eq. (4.30) is satisfied in the case of infinite water depth in the cases of both partial and complete barriers, the details of which are left as exercises.

Further, gravity wave scattering by double flexible porous barriers can be treated in a similar manner by the suitable application of expansion formulae. In addition, gravity wave scattering by flexible barriers in two-layer fluids having free surface and interface are left in the context of the present monograph in the case of infinite water depths which can be attempted utilising the expansion formulae for two-layer fluid as discussed in Section 2.6.

4.5 Method of wide spacing approximation

The method of wide spacing approximation is applied to varieties of physical problems associated with wave scattering by multiple objects (see [32], [60], [72] and [114]). In the context of the present monograph, the wide spacing approximation method is briefly discussed here for computing the reflection and transmission coefficients associated with wave scattering by multiple flexible barriers of similar configuration and characteristics from the known values of reflection and transmission coefficients associated with single barrier.

The scattering of surface water waves by multiple barriers will be discussed in the case of finite water depth here assuming that the fluid characteristics are the same as discussed in Section 4.2 and all the flexible porous barriers are of the same characteristics and configurations as discussed in Section 4.2. It is assumed that the N flexible porous barriers fully extended from the free surface to the bottom are located at $x = a_j$ for $j = 1, 2, ..., N$. In the present section, the objective is to determine the reflection and transmission coefficients as the waves transmit through the N flexible porous barriers. It is assumed that a monochromatic gravity wave is normally incident from the negative x direction on the first barrier at $x = a_1$ and passed through the multiple porous barriers. It experiences partial reflection and transmission by each barrier before being transmitted into the open water region. In this case, N flexible barriers occupy the regions $-\infty < x < a_1$, $a_j < x < a_{j+1}$ for $j = 1, 2, ..., N$ and $a_n < x < \infty$ along the x-axis and $0 < y < h$ along the y-axis. Assuming that the fluid is inviscid and incompressible and the motion is irrotational and simple harmonic in time with angular frequency ω, the velocity potential $\Phi_j(x, y, t)$ is expressed in the form $\Phi_j(x, y, t) = Re\{\phi_j(x, y)e^{-i\omega t}\}$ where Re denotes the real part and the subscript j refers to the respective regions. The spatial velocity potential $\phi_j(x, y)$ satisfies the governing Laplace equation as given by

$$\nabla^2 \phi_j = 0 \quad \text{on} \quad 0 < y < h, \quad -\infty < x < \infty. \tag{4.131}$$

The linearised plate-covered boundary condition on the mean free surface is given by

$$(\partial_y + K)\phi_j = 0 \quad \text{on} \quad y = 0, \quad x \in I_j, \quad j = 1, ..., N + 1, \qquad (4.132)$$

where $K = \omega^2/g$. It may be noted that $I_1 = (-\infty, a_1), I_{j+1} = (a_j, a_{j+1}), j = 1, 2, ..., N - 1$ and $I_{N+1} = (a_N, \infty)$. The rigid bottom boundary condition yields

$$\partial_y \phi_j = 0 \quad \text{on} \quad y = h. \qquad (4.133)$$

Proceeding in a similar manner as in Eqs. (4.8)–(4.13), the boundary condition on the flexible breakwaters is given by

$$\frac{\partial \phi_j}{\partial x} = ik_0 G\{\phi_j(x, y) - \phi_{j+1}(x, y)\} - i\omega\xi_j, \text{ at } x = a_j, \; j = 1, 2, ..., N, \; (4.134)$$

with $\xi_j(y)$ satisfying the equation

$$EI\frac{d^4\xi_j}{dy^4} - Q\frac{d^2\xi_j}{dy^2} - m_s\omega^2\xi_j = i\omega\rho(\phi_j - \phi_{j+1}), \text{ at } x = a_j, \; j = 1, 2, ..., N.$$
$$(4.135)$$

In addition, across the flexible porous barriers, the continuity of velocity for $j = 1, ..., N$ yields

$$\phi_{jx}(x+, y) = \phi_{(j+1)x}(x-, y), \text{ at } x = a_j \text{ for all } y. \qquad (4.136)$$

The far field radiation condition is given by

$$\phi_1(x, y) \sim (I_0 e^{ik_0 x} + R_{N0} e^{-ik_0 x})\psi_0(y) \quad \text{as} \quad x \to -\infty,$$
$$\phi_{N+1}(x, y) \sim T_{N0} e^{ik_0 x} \psi_0(y) \qquad \qquad \text{as} \quad x \to \infty, \qquad (4.137)$$

where $\psi_0(y)$ is the same as defined in Eq. (4.6), with k_0 satisfying the gravity wave dispersion relation given in Eq. (4.7) and I_0 being the amplitude of the incident waves assumed to be known. The unknown constants R_{N0} and T_{N0} are associated with the amplitude of the reflected and transmitted waves in the case of N barriers. Similar to Eq. (4.14), assuming that the flexible structures are fixed at seabed and are free at the free surface, the boundary conditions at the two ends of the structure lead to the vanishing of the bending moment and shear force near the free surface and the vanishing of the structural deflection and slope of deflection at the bottom bed of the structure which yield

$$\xi_j(h) = 0, \; \xi_{jy}(h) = 0, \; \xi_{jyy}(0) = 0, \; \xi_{jyyy}(0) = 0, \; j = 1, 2, ..., N, \qquad (4.138)$$

while, in the case of infinite water depth, the barrier is assumed to be fixed at the free surface, which leads to the conditions

$$\xi_j(0) = 0, \; \xi_{jy}(0) = 0, \quad \text{and} \quad \xi_j, \xi_{jy} \to 0 \quad \text{as} \quad y \to \infty. \qquad (4.139)$$

Thus, the velocity potentials $\phi_j(x,y)$ in each of the $(N+1)$ regions in the case of finite water depth are given by

$$
\phi_j = \begin{cases}
I_0 e^{ik_0(x-a_1)} \psi_0(y) + \displaystyle\sum_{n=0}^{\infty} R_{Nn} e^{k_n(x-a_1)} \psi_n(y), & x < a_1, \\[3mm]
\displaystyle\sum_{n=0}^{\infty} \left\{ A_n^j \frac{\cosh k_n x}{\cosh k_n \delta a_j} + B_n^j \frac{\sinh k_n x}{\sinh k_n \delta a_j} \right\} \psi_n(y), & x \in \displaystyle\bigcup_{j=2}^{N} I_j, \\[3mm]
\displaystyle\sum_{n=0}^{\infty} T_{Nn} e^{-k_n(x-a_N)} \psi_n(y), & x > a_N,
\end{cases} \tag{4.140}
$$

where $k_n = -ik_0$ for $n = 0$. The eigenfunctions $\psi_n(y)$ are the same as in Eq. (4.16) with the eigenvalues p_n satisfying the dispersion relation given in Eq. (4.7) and satisfying the orthogonal relation in Eq. (4.17). On the other hand, the velocity potentials $\phi_j(x,y)$ in each of the $(N+1)$ regions in the case of infinite water depth are expressed in terms of appropriate eigenfunctions as given by

$$
\phi_j = \begin{cases}
a_0(x)\psi_0(y) + \dfrac{2}{\pi} \displaystyle\int_0^{\infty} \dfrac{L(\xi,y)R(\xi)e^{(x-a_1)\xi}d\xi}{\Delta(\xi)}, & x < a_1, \\[4mm]
a_j(x)\psi_0(y) + \dfrac{2}{\pi} \displaystyle\int_0^{\infty} \dfrac{C_j(x,\xi)L(\xi,y)d\xi}{\Delta(\xi)}, & x \in \displaystyle\bigcup_{j=2}^{N} I_j, \\[4mm]
a_{N+1}(x)\psi_0(y) + \dfrac{2}{\pi} \displaystyle\int_0^{\infty} \dfrac{T(\xi)L(\xi,y)e^{-\xi(x-a_N)}d\xi}{\Delta(\xi)}, & x > a_N,
\end{cases} \tag{4.141}
$$

where $a_0(x) = I_0 e^{ik_0(x-a_1)} + R_{N0} e^{-ik_0(x-a_1)}$, $a_j(x) = A_j \cos k_0 x + B_j \sin k_0 x$, $a_{N+1}(x) = T_{N0} e^{ik_0(x-a_N)}$, $C_j(x,\xi) = A_j(\xi)\cosh \xi x + B_j(\xi)\sinh \xi x$ with I_js being the same as defined in case of finite water depth. The eigenfunctions $\psi_0(y)$ and $L(\xi,y)$ are the same as in Eq. (4.83) and satisfy the orthogonal mode-coupling relation in Eq. (4.84). Using the conditions on the breakwater as in Eqs. (4.91)–(4.93) and the edge conditions in (4.95), the boundary value problems can be converted to (i) a system of equations in the case of water of finite depth and (ii) a system of integral equations in the case of water of infinite depth from which all the unknowns associated with the velocity potential can be obtained. Once the unknowns R_{N0} and T_{N0} are obtained, the reflection and transmission coefficients are derived from the relations $K_r = |R_{N0}/I_0|$ and $K_t = |T_{N0}/I_0|$. It may be further noted that K_r and K_t satisfy the energy relation $K_r^2 + K_t^2 = 1$, which can be used to check the computed numerical results. The details are deferred here. However, to determine the reflection and transmission coefficients associated with wave scattering by multiple barriers as discussed above, the wide spacing approximation method can be used which is briefly demonstrated here.

The method of wide spacing approximation is based on the assumption that the distance between two consecutive barriers is much larger than the wave length of the incident plane progressive wave, that is, $|a_{j+1} - a_j| \gg \lambda$, for $j = 1, 2, ..., N - 1$ with λ being the incident wave length (as in [29], [92]) to ensure that the evanescent modes do not contribute to the solution. Thus, the local effects produced during the interaction of the incident wave with one of the barriers do not affect the subsequent interactions. In the case of wave scattering by multiple flexible porous barriers, it may be noted that the surface gravity wave experiences partial reflection and transmission by each of the flexible porous barriers located at $x = a_j$ for $j = 1, 2, ..., N$. Thus, using Eq. (4.134), the asymptotic forms of the velocity potentials ϕ_j far away from the barriers in the respective regions are given by

$$
\begin{aligned}
\phi_1 &\sim (I_0 e^{ik_0 x} + R_{N0} e^{-ik_0 x}) f_0(y), \quad -\infty < x < a_1, \\
\phi_{j+1} &\sim (A_j e^{ik_0 x} + B_j e^{-ik_0 x}) f_0(y), \quad j = 1, 2, ..., N - 1, \\
\phi_{N+1} &\sim T_{N0} e^{ik_0 x} f_0(y), \quad a_N < x < \infty,
\end{aligned}
\tag{4.142}
$$

with $a_j < x < a_{j+1}$ for $j = 1, 2, ..., N - 1$. Equating the left and right going components of the propagating waves at the barrier location $x = a_j$ for $j = 1, 2, ..., N$ with the amplitudes of the reflected and transmitted waves in the prescribed region, a system of $2N$ linear equations associated with $2N$ unknowns $R_{N0}, T_{N0}, A_{j0}, B_{j0}, j = 1, 2, ..., N - 1$ is obtained as given by

$$
\begin{aligned}
R_{N0} e^{ik_0 a_1} &= R_{10} e^{-ik_0 a_1} + B_1 T_{10} e^{ik_0 a_1}, \\
A_j e^{ik_0 a_j} &= T_{10} A_{j-1} e^{-ik_0 a_j} + B_j R_{10} e^{ik_0 a_j}, \\
B_j e^{ik_0 a_{j+1}} &= A_j R_{10} e^{-ik_0 a_{j+1}} + T_{10} B_{j+1} e^{ik_0 a_{j+1}}, \\
T_{N0} e^{-ik_0 a_N} &= A_{N-1} T_{10} e^{-ik_0 a_N},
\end{aligned}
\tag{4.143}
$$

with $A_0 = I_0$ and $B_N = 0$. Here R_{10} and T_{10} correspond to the amplitudes of the reflected and transmitted waves by a single barrier located at $x = a_j$, $j = 1, 2, ..., N$ in isolation. Thus, by solving the above system of equations, the reflection and transmission coefficients $K_r = |R_{N0}/I_0|$ and $K_t = |T_{N0}/I_0|$ for N barriers are obtained.

4.6   Conclusion

In the present chapter, various wave structure interaction problems are discussed in which the structure is modeled as a beam and is under uniform axial force. In the discussed physical problems, the edge conditions are assumed to be fixed or free. However, problems with different types of edge conditions can be suitably incorporated at the edges to analyse problems of practical interest. The same modeling concept and methodology can be easily generalised to deal with wave interactions with a flexible porous membrane in a straightforward manner with appropriate edge conditions. In the present chapter, emphasis is on the eigenfunction expansion method and the least square approximation method. Another important method which is widely used in the literature is the boundary integral equation method. This method is based on the application of the free surface Green's function, which has the advantage that it will be suitable for arbitrary geometry apart from wave scattering by partial/complete barriers (for details see [17], [70] and [143]). Most of these results are discussed for wave interactions with flexible vertical porous/non-porous structures associated with normalized or obliquely incident waves in the Cartesian coordinate system. There is significant progress on wave interaction with flexible porous/nonporous cylinders (see [155]). Surface wave scattering by flexible porous cylinders can be analysed with the help of the Fourier-Bessel series solution in the case of finite water depth and appropriate integral transforms involving the Bessel function and Fourier transforms in the case of infinite water depth. Among other methods which can be used to solve this class of problems are the Wiener-Hopf technique and the method of residue calculus. Further, all the problems are analysed based on small amplitude water wave theory and structural response. However, there is negligible attempt in the literature to deal with wave interaction with vertical structures under the assumption of long wave theory and finite amplitude wave theory. In the present chapter, all the problems are analysed assuming that the bottom bed is uniform in nature. However, small variation in bottom topography can be taken into consideration to analyse wave interaction with flexible structures. Further, the modeling concept can be easily used to reduce the resonating effect in wave tanks and basins with the help of multiple flexible porous screens. Another major aspect of these problems is to incorporate the singular behavior of the velocity potential at the tip in the case of a partial flexible porous barrier. Further, the surface wave scattering and trapping by flexible porous barriers of different configurations can be attempted for solution in infinite water depth in the cases of both single- and two-layer fluids, the details of which are deferred in the present monograph.

4.7 Examples and exercises

Exercise 4.1 Using the free surface Green's function, discuss the solution procedure to derive the velocity potentials associated with the scattering of surface gravity waves by a vertical flexible porous barrier (for the cases of both complete and partial barriers) in water of finite and infinite water depths.

Exercise 4.2 Using the free surface Green's function, discuss the solution procedure to derive the velocity potentials associated with the trapping of surface gravity waves by a vertical flexible porous barrier (for the cases of both complete and partial barriers) in water of finite and infinite water depths.

Exercise 4.3 Using the free surface Green's function, discuss the solution procedure to derive the velocity potentials associated with the scattering of surface gravity waves by multiple vertical flexible porous barriers (for the cases of both complete and partial barriers) in water of finite and infinite water depths.

Exercise 4.4 Using the free surface Green's function, discuss the solution procedure to derive the velocity potentials associated with wave trapping by a flexible porous barrier near a rigid wall (for the cases of both complete and partial barriers) in water of finite and infinite water depths.

Exercise 4.5 Using the appropriate Green's function, discuss the solution procedure to derive the velocity potentials associated with the scattering of surface gravity waves by a vertical flexible porous barrier (for the cases of both complete and partial barriers) in water of finite and infinite water depths in a two-layer fluid having a free surface and an interface.

Exercise 4.6 Using the free surface Green's function for two-layer fluids having a free surface and interface, discuss the solution procedure to derive the velocity potentials associated with wave trapping by a flexible porous barrier near a rigid wall (for the cases of both complete and partial barriers) in water of finite and infinite water depths in a two-layer fluid having free surface and interface.

Exercise 4.7 Discuss the solution procedure to determine the velocity potential associated with the scattering of surface gravity waves by a porous flexible cylinder as in Figure 4.8.

Exercise 4.8 Discuss the solution procedure to determine the velocity potential associated with the trapping of surface gravity waves by two concentric

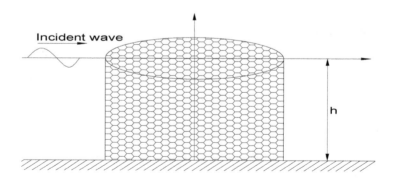

FIGURE 4.8
Schematic diagram of wave scattering by vertical flexible porous cylinder.

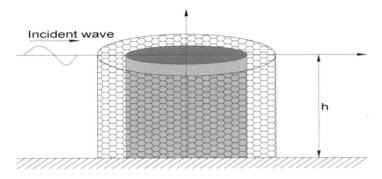

FIGURE 4.9
Schematic diagram of wave trapping by two concentric cylinders.

cylinders of different radii assuming that the inner cylinder is rigid and the outer cylinder is flexible and porous as in Figure 4.9. It may be assumed that in the case of the flexible cylinder, the cylinder is fixed at the bottom bed and has a free edge near the free surface.

Exercise 4.9 Discuss the solution procedure to determine the velocity potential associated with the trapping of surface gravity waves by two concentric cylinders of different radii assuming that the inner cylinder is flexible and porous and the outer cylinder is rigid. It may be assumed that in the case of the flexible cylinder, the cylinder is fixed at the bottom bed and has a free edge near the free surface. Discuss the role of such a system in reducing wave oscillation in a liquid-filled tank having a free surface.

Exercise 4.10 Discuss the solution procedure to determine the velocity potential associated with the scattering of surface gravity waves by a flexible porous truncated cylinder in water of finite depth as in Figure 4.10. It may be assumed that in the case of the flexible cylinder, the cylinder is fixed at both the ends.

Exercise 4.11 Discuss the solution procedure to determine the velocity potential associated with the trapping of surface gravity waves by two concentric truncated cylinders of different radii assuming that the inner cylinder is rigid and the outer cylinder is flexible and porous as in Figure 4.11. It may be assumed that in the case of the flexible cylinder, the cylinder is fixed both at the bottom bed and at the free surface.

Exercise 4.12 Discuss the solution procedure to determine the velocity potential associated with the scattering of surface gravity waves by a flexible floating circular net of radius a used for fish farming in water of finite depth. It may be assumed that the net is made up of a membrane which is under uniform tension T and is kept fixed near the free surface and at the bottom bed.

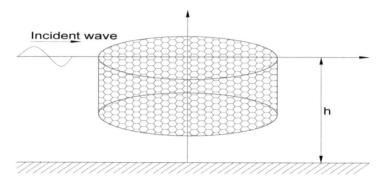

FIGURE 4.10
Schematic diagram of wave scattering by truncated flexible porous cylinder.

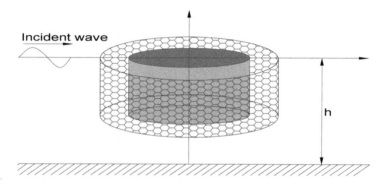

FIGURE 4.11
Schematic diagram of wave trapping by two concentric particle cylinders.

Exercise 4.13 Discuss the solution procedure to determine the velocity potential associated with the scattering of gravity waves in surface and internal modes by a partial flexible floating porous barrier in water of finite and infinite water depths.

Exercise 4.14 Discuss the solution procedure to determine the velocity potential associated with the trapping of gravity waves in surface and internal modes by a partial flexible floating porous barrier kept near a rigid wall in water of finite and infinite water depths.

5

Time domain analysis of wave structure interaction problems

5.1 General introduction

In the previous chapters, all the physical problems are considered for solution based on the assumption that the motion is simple harmonic in time, which has reduced the physical problems to boundary value problems. However, in the presence of a sudden disturbance such as an impulse or in the case of a vehicle moving on the surface of a floating elastic plate or during landing and taking off of an aircraft on a floating plate-/ice-covered free surface, this simple harmonic motion is not a valid assumption. Further, often there are situations where the disturbances grow with the increase in time assuming that there was no disturbance at time $t = 0$. The significance of the associated wave structure interaction problem is that t is not an active variable of the governing differential equation. However, the time derivative appears on the plate-covered mean free surface and elsewhere with higher order derivative terms arising on the structural boundaries. To deal with such time-dependent problems, often the Laplace transform is employed to reduce the initial boundary value problem to a boundary value problem of the type discussed in the previous chapters. However, in many situations for problems in the unbounded domain, an appropriate integral transform method is employed (Hankel transform is used for axi-symmetric problems) to obtain integral representation of the boundary value problems incorporating all the boundary conditions, and asymptotic results are obtained using the method of stationary phase. In many other situations, Green's integral theorem is employed along with a suitable Green's function to deal with the time domain problem in a direct manner. There has been significant progress on initial value problems for transient water waves since Finkelstein [35]. Recently, there is a growing interest in dealing with time domain unsteady wave structure interaction problems due to their wide applications in various areas of marine science and technology including polar science (see [67], [124] and other cited papers). Significant progress on time-dependent wave structure interaction problems has been based on the mode expansion method, the transform method and the spectral method (as in [36], [65], [96], [98], [132], [135], and the cited literature). There are several studies for determining the source singularities associated with unsteady water wave motion in single- and double-layer fluid having free surface and interface (see [108] and the literature cited therein). Due to the growing interest in the wave structure interaction problem, attempts are being made to generalise several results available in the literature for gravity wave problems to apply to flexural gravity wave motion problem as in [11] and [83].

5.2   Mathematical formulation

The problem of the generation of flexural gravity waves associated with the interaction of surface gravity waves with a floating ice sheet which is modeled as an elastic plate is considered assuming that the motion is small and commences at time $t = 0$ from a state of rest. The floating ice sheet occupies the region $-\infty < x < \infty, y = 0$. The problem is considered in the two-dimensional Cartesian coordinate system (x, y) with the x-axis being taken in the horizontal direction and y-axis being taken in the vertically downward positive direction with the fluid occupying the region $-\infty < x < \infty, 0 < y < h$ in the case of water of finite depth h and $-\infty < x < \infty, 0 < y < \infty$ in the case of infinite water depth. The ice sheet of small thickness d is floating on the undisturbed water surface $-\infty < x < \infty, y = 0$. In the linearised theory of water waves as assumed in Chapter 1, the fluid is assumed to be inviscid and incompressible and the motion is irrotational in nature. Thus, the associated velocity potential $\Phi(x, y, t)$ satisfies the PDE

$$\nabla^2 \Phi = 0, \quad \text{in the fluid domain,} \tag{5.1}$$

along with the linearised kinematic and the dynamic boundary conditions at the ice-covered mean free surface $y = 0$ as given by

$$\frac{\partial \Phi}{\partial y} = \frac{\partial \eta}{\partial t}, \tag{5.2}$$

$$\frac{1}{g}\frac{\partial \Phi}{\partial t} = \left(D\frac{\partial^4}{\partial x^4} + Q\frac{\partial^2}{\partial x^2} + \gamma\frac{\partial^2}{\partial t^2} + 1\right)\eta, \tag{5.3}$$

where $\eta(x,t)$ = vertical plate deflection, $\gamma = \rho_i d/\rho g$, $D = EI/\rho g$, $I = d^3/12(1-\nu^2)$, $Q = N/\rho g$, ρ = water density, ρ_i = ice density, E = Young's modulus, N = axial force, d = plate thickness, g = acceleration due to gravity and ν = the Poisson ratio. The no flow condition at the bottom boundary yields

$$\frac{\partial \Phi}{\partial y} = 0, \quad \text{on } y = h \quad \text{in the case of finite water depth} \tag{5.4}$$

and

$$\Phi, |\nabla\Phi| \to 0, \quad \text{as } y \to \infty \quad \text{in the case of infinite water depth.} \tag{5.5}$$

The initial conditions are

$$\Phi(x,0,0) = 0, \quad \eta(x,0) = 0, \quad \eta_t(x,0) = 0. \tag{5.6}$$

Combining the kinematic and the dynamic upper surface boundary conditions (5.2) and (5.3), the linearised ice-covered surface boundary condition is obtained as

$$\left(D\frac{\partial^4}{\partial x^4} + Q\frac{\partial^2}{\partial x^2} + \gamma\frac{\partial^2}{\partial t^2} + 1\right)\frac{\partial\Phi}{\partial y} = \frac{1}{g}\frac{\partial^2\Phi}{\partial t^2}. \tag{5.7}$$

Here, it may be noted that the boundary condition on the ice-/plate-covered boundary is of higher order than the governing equation which makes the problem more complex. Further, the speciality of these problems is that the time variable t is not an active variable of the governing equation which is the Laplace equation. Define the Laplace transform of a function $f(t)$ as given by

$$\bar{f}(p) = \int_0^\infty f(t)e^{-pt}dt, \tag{5.8}$$

where p is the Laplace transform variable and $\bar{f}(p)$ refers to the Laplace transform of $f(t)$. Applying the Laplace transform in time variable t as defined above, Eq. (5.1) yields

$$\nabla^2\bar{\Phi} = 0, \quad \text{in the fluid domain.} \tag{5.9}$$

Using the initial conditions in Eq. (5.6), the Laplace transform of the kinematic condition in Eq. (5.2) yields

$$\frac{\partial\bar{\Phi}}{\partial y} = p\bar{\eta}(x,p), \quad \text{on } y = 0. \tag{5.10}$$

Further, using the initial conditions in Eq. (5.6) the Laplace transform of the linearised plate-covered boundary condition in Eq. (5.7) yields

$$\left(D\frac{\partial^4}{\partial x^4} + Q\frac{\partial^2}{\partial x^2} + \gamma p^2 + 1\right)\frac{\partial\bar{\Phi}}{\partial y} = \frac{p^2}{g}\bar{\Phi}, \quad \text{on } y = 0. \tag{5.11}$$

In addition, $\bar{\Phi}$ satisfies the bottom boundary conditions as in Eqs. (5.4) and (5.5). Thus, the initial boundary value problem in $\Phi(x,y,t)$ is converted into a boundary value problem in $\bar{\Phi}(x,y,p)$, which is similar to the boundary value problem associated with the flexural gravity wave motion. The boundary value problem in $\bar{\Phi}$ can be attempted for a solution based on the suitable application of Green's function method or the Fourier transform technique, the details of which are discussed in specific physical problems in the later sections.

Hereafter in the present chapter, it is assumed that the thickness of the floating ice sheet is small compared to the wavelength of the flexural gravity waves (i.e., $\gamma k \ll 1$) so that the term containing γ is neglected as in previous chapters.

5.3   Unsteady wave structure interaction problems

In this section, using the Laplace transform, the time-dependent initial value problems associated with wave structure interaction problems are converted into boundary value problems, which are similar to the time-harmonic boundary value problems. The solutions of the associated boundary value problems are obtained by suitable generalisations of the methods applied to time-harmonic problems.

5.3.1   Time-dependent Green's functions

The time-dependent Green's function $G(x, y; x_0, y_0, t)$ dealing with surface wave interaction with floating elastic plate is derived assuming that the source is of strength $m(t)$ in the cases of water of both finite and infinite depths as discussed in Section 5.2. Green's function $G(x, y, x_0, y_0, t)$ satisfies the Laplace equation

$$\frac{\partial^2 G}{\partial x^2} + \frac{\partial^2 G}{\partial y^2} = 0, \tag{5.12}$$

in the fluid domain except at (x_0, y_0) along with the free surface boundary condition as given by the linearised ice-/plate-covered boundary condition on the mean free surface in the presence of compressive force as in Eq. (5.7), and is given by

$$D\frac{\partial^5 G}{\partial y^5} - Q\frac{\partial^3 G}{\partial y^3} + \frac{\partial G}{\partial y} - \frac{1}{g}\frac{\partial^2 G}{\partial t^2} = 0, \quad \text{on} \quad y = 0, \quad 0 < x < \infty. \tag{5.13}$$

Green's function $G(x, y, x_0, y_0, t)$ satisfies the bottom boundary condition (5.4) or (5.5) as appropriate for water of finite or infinite depths. In addition, $G(x, y, x_0, y_0, t)$ satisfies the condition

$$G(x, y, x_0, y_0, t) \sim m(t)\ln r, \quad \text{as} \quad r = \sqrt{(x - x_0)^2 + (y - y_0)^2} \to 0, \tag{5.14}$$

where $m(t)$ is the strength of the source located at (x_0, y_0). Further, $G(x, y, x_0, y_0, t)$ satisfies

$$|G|, |\nabla G| \to 0, \quad \text{as} \quad r = \sqrt{(x - x_0)^2 + (y - y_0)^2} \to \infty \tag{5.15}$$

and

$$G, G_t = 0, \quad \text{at} \quad t = 0, \quad \text{on} \quad y = 0. \tag{5.16}$$

The determination of Green's function G based on the direct application of the Laplace transform as in [40] and [116] is demonstrated in brief. Taking the Laplace transform of Eq. (5.12) satisfying the boundary conditions in Eqs. (5.13)–(5.16) and the bottom conditions in Eqs. (5.4) and (5.5), it can be easily derived that the transformed function $\bar{G}$ satisfies Eq. (5.12) and the boundary conditions in Eqs. (5.4) and (5.5). The free surface condition in Eq. (5.3) reduces to

$$D\frac{\partial^5 \bar{G}}{\partial y^5} - Q\frac{\partial^3 \bar{G}}{\partial y^3} + \frac{\partial \bar{G}}{\partial y} - \frac{p^2}{g}\bar{G} = 0, \quad \text{on} \quad y = 0, \quad 0 < x < \infty. \quad (5.17)$$

Further, from Eq. (5.14), it can be easily derived that

$$\bar{G}(x, y, x_0, y_0, p) \sim \bar{m}(p)\ln r, \quad \text{as} \quad r = \sqrt{(x - x_0)^2 + (y - y_0)^2} \to 0. \quad (5.18)$$

Next, the transformed Green's function $\bar{G}(x, y, x_0, y_0, p)$ in water of infinite and finite water depths will be derived independently as in [87]. Keeping the condition (5.18) in mind, the transformed Green's function in the case of infinite water depth is expanded in the form given by

$$\bar{G}(x, y, x_0, y_0, p) = \bar{m}(p)\left\{\ln\frac{r}{r'} + \int_0^\infty A(\xi)e^{-\xi y}\cos\xi(x - x_0)d\xi\right\}, \quad (5.19)$$

where $r' = \sqrt{(x - x_0)^2 + (y + y_0)^2}$ and $0 < y_0 < \infty$, so that the singularity is submerged but not on the bottom in infinite water depth. Using the plate-covered surface boundary condition (5.17) and results (see [40] and [137])

$$\int_0^\infty \xi^{2n}e^{-\xi|x-x_0|}\sin\xi y_0 d\xi = (-1)^n \int_0^\infty \xi^{2n}e^{-\xi y_0}\cos\xi|x - x_0|d\xi, \quad (5.20)$$

and

$$\ln\frac{r}{r'} = \begin{cases} -2\displaystyle\int_0^\infty \frac{e^{-\xi y}}{\xi}\sinh\xi y_0\cos\xi(x - x_0)d\xi & \text{for} \quad y > y_0, \\[4mm] -2\displaystyle\int_0^\infty \frac{e^{-\xi y_0}}{\xi}\sinh\xi y\cos\xi(x - x_0)d\xi & \text{for} \quad y < y_0, \end{cases} \quad (5.21)$$

it can be easily derived that

$$A(\xi) = -\frac{2g(1 - Q\xi^2 + D\xi^4)e^{-\xi y_0}}{g\xi(1 - Q\xi^2 + D\xi^4) + p^2}. \quad (5.22)$$

Hence, for $|x - x_0| > 0$, Green's function is expressed as

$$\bar{G} = \bar{m}(p)\left\{\ln\frac{r}{r'} - 2\int_0^\infty \frac{g(1 - Q\xi^2 + D\xi^4)e^{-\xi(y+y_0)}}{g\xi(1 - Q\xi^2 + D\xi^4) + p^2}\cos\xi|x - x_0|d\xi\right\}. \quad (5.23)$$

Using the convolution formula and taking the Laplace inversion, Eq. (5.23) yields

$$G = m(t)\ln\frac{r}{r'} - 2\int_0^\infty \frac{\Omega e^{-\xi(y+y_0)}\cos\xi|x - x_0|}{\xi}\int_0^t m(\tau)\sin\Omega(t - \tau)d\tau d\xi, \quad (5.24)$$

where $\Omega = \sqrt{g\xi(1 - Q\xi^2 + D\xi^4)}$. Thus, $G(x, y, x_0, y_0, t)$ in Eq. (5.24) represents the time domain Green's function with strength $m(t)$ in infinite water depth.

Proceeding in a similar way as in the case of infinite depth, the time domain Green's function with strength $m(t)$ in the case of finite depth is expanded into the form

$$\bar{G} = \bar{m}(p)\left[\ln\frac{r}{r'} + \int_0^\infty \{A(\xi)\cosh\xi(h - y) + B(\xi)\sinh\xi y\}\cos\xi|x - x_0|d\xi\right],$$

(5.25)

with $0 < y_0 < h$, so that the singularity is submerged but not on the bottom. Using the boundary conditions in Eqs. (5.4) and (5.17), $A(\xi)$ and $B(\xi)$ are obtained as

$$A(\xi) = \frac{-g(1 - Q\xi^2 + D\xi^4)}{g\xi(1 - Q\xi^2 + D\xi^4)\sinh\xi h + p^2\cosh\xi h}\frac{\cosh\xi(h - y_0)}{\cosh\xi h}, \quad (5.26)$$

$$B(\xi) = \frac{-e^{\xi h}\sinh\xi y_0}{\xi\cosh\xi h}. \quad (5.27)$$

Hence, from Eqs. (5.25)–(5.27), Green's function $\bar{G}$ is obtained as

$$\bar{G}(x, y; x_0, y_0, p) = \bar{m}(p)\left[\ln\frac{r}{r'} - 2\int_0^\infty \frac{e^{-\xi h}g_1(y, y_0)\cos\xi(x - x_0)}{\xi\cosh\xi h}d\xi\right.$$

$$\left. - 2\int_0^\infty \frac{\Omega_h^2 g_2(y, y_0)}{\Omega_h^2 + p^2}\frac{\cos\xi(x - x_0)}{\xi\sinh\xi h\cosh\xi h}d\xi\right], \quad (5.28)$$

where $\Omega^2 = g\xi(D\xi^4 - Q\xi^2 + 1)\tanh\xi h$, $g_1(y, y_0) = \sinh\xi y\sinh\xi y_0$ and $g_2(y, y_0) = \cosh\xi(h - y)\cosh\xi(h - y_0)$. Using the convolution theorem and taking the inverse Laplace transform of $\bar{G}$, Eq. (5.28) yields

$$G = m(t)\ln\frac{r}{r'} - 2\left[m(t)\int_0^\infty \frac{e^{-\xi h}g_1(y, y_0)\cos\xi(x - x_0)}{\xi\cosh\xi h}d\xi\right.$$

$$\left. - \int_0^\infty \frac{\Omega_h(\xi)g_2(y, y_0)\cos\xi(x - x_0)}{\xi\sinh\xi h\cosh\xi h}\left\{\int_0^t m(\tau)\sin\Omega_h(t - \tau)d\tau\right\}d\xi\right].$$

(5.29)

Thus, Green's functions for the time domain problem are derived in the cases of both finite and infinite water depths. Green's functions G given in Eqs. (5.24) and (5.29) are often called the velocity potentials due to a source of strength $m(t)$ located at (x_0, y_0) associated with flexural gravity wave motion in water of infinite and finite depths, respectively. Further, from Eqs. (5.2) and (5.10), using the convolution theorem and taking the Laplace inversion, the plate deflection due to a source potential of strength $m(t)$ can be obtained as

$$\eta(x,t) = \int_0^t \{G_y(\tau)\}_{y=0} d\tau. \tag{5.30}$$

Hence, from Eqs. (5.24) and (5.30), the plate deflection $\eta(x,t)$ in the case of a source of strength $m(t)$ in the case of infinite water depth is derived as

$$\eta(x,t) = -2 \int_0^\infty e^{-ky_0} \cos k(x-x_0) \int_0^t m(\tau) \cos \Omega(t-\tau) d\tau dk. \tag{5.31}$$

In a similar manner, from Eqs. (5.29) and (5.30), the plate deflection $\eta(x,t)$ due to a source potential of strength $m(t)$ in the case of finite water depth can be derived and details are left as an exercise.

As a special case of the general derivation of source potential of strength $m(t)$, when the wave source has impulse strength $m(t) = \delta(t)$, from Eq. (5.24) the corresponding source potential G_δ in the case of infinite water depth is obtained as

$$G_\delta = \delta(t) \ln \frac{r}{r'} - 2 \int_0^\infty \frac{\Omega e^{-\xi(y+y_0)} \cos \xi |x-x_0|}{\xi} \sin \Omega t d\xi, \tag{5.32}$$

while from Eq. (5.29), the wave source potential in finite water depth is obtained as

$$G = \delta(t) \ln \frac{r}{r'} - 2 \left[\delta(t) \int_0^\infty \frac{e^{-\xi h} g_1(y,y_0) \cos \xi(x-x_0)}{\xi \cosh \xi h} d\xi \right.$$
$$\left. - \int_0^\infty \frac{\Omega_h(\xi) g_2(y,y_0) \cos \xi(x-x_0)}{\xi \sinh \xi h \cosh \xi h} \sin \Omega_h t d\xi \right]. \tag{5.33}$$

Further, the wave source potential of constant strength $m = 1/2\pi$ in the case of infinite water depth is given by

$$G = \frac{1}{2\pi} \ln \frac{r}{r'} - \frac{1}{\pi} \int_0^\infty \frac{\Omega e^{-\xi(y+y_0)} \cos \xi |x-x_0|}{\xi} \int_0^t \sin \Omega(t-\tau) d\tau d\xi, \tag{5.34}$$

while in the case of finite water depth, the wave source potential of constant strength $m = 1/2\pi$ is given by

$$G = \frac{1}{2\pi} \ln \frac{r}{r'} - \frac{1}{\pi} \left[\int_0^\infty \frac{e^{-\xi h} g_1(y,y_0) \cos \xi |x-x_0|}{\xi \cosh \xi h} d\xi \right.$$
$$\left. - \int_0^\infty \frac{\Omega_h(\xi) g_2(y,y_0) \cos \xi |x-x_0|}{\xi \sinh \xi h \cosh \xi h} \left\{ \int_0^t \sin \Omega_h(t-\tau) d\tau \right\} d\xi \right]. \tag{5.35}$$

Further, from the Green's function $\bar{G}$ derived in Eq. (5.21), the solution of the time-harmonic Green's function $G(x,y,x_0,y_0;t)$ of the form $G(x,y,x_0,y_0;t) =$

$Re\{\tilde{G}(x, y, x_0, y_0)e^{-i\omega t}\}$ can be derived in a straightforward manner, as discussed briefly below. It is easy to observe that $\tilde{G}(x, y, x_0, y_0)$ satisfies the Laplace equation in Eq. (5.10) along with the bottom boundary conditions in Eqs. (5.4) and (5.5) as appropriate in the case of water of finite and infinite depths. Further, $\tilde{G}(x, y, x_0, y_0)$ satisfies the free surface boundary condition

$$D\frac{\partial^5 \tilde{G}}{\partial y^5} - Q\frac{\partial^3 \tilde{G}}{\partial y^3} + \frac{\partial \tilde{G}}{\partial y} - K\tilde{G} = 0, \quad \text{on} \quad y = 0, \quad 0 < x < \infty, \qquad (5.36)$$

where $K = \omega^2/g$. In addition, $\tilde{G}(x, y; x_0, y_0)$ satisfies the condition

$$\tilde{G}(x, y, x_0, y_0) \sim \frac{1}{2\pi}\ln r, \quad \text{as} \quad r = \sqrt{(x - x_0)^2 + (y - y_0)^2} \to 0. \qquad (5.37)$$

It is easy to check that $\tilde{G}$ satisfies the governing equation and conditions satisfied by $\bar{G}$ with $K = -p^2$ and $\bar{m}(p) = 1/2\pi$. Thus, Eq. (5.23) yields

$$\tilde{G} = \frac{1}{2\pi}\ln\frac{r}{r'} - \frac{1}{\pi}\int_0^\infty \frac{(1 - Q\xi^2 + D\xi^4)e^{-\xi(y+y_0)}}{\xi(1 - Q\xi^2 + D\xi^4) - K}\cos\xi|x - x_0|d\xi. \qquad (5.38)$$

Using the Cauchy integral theorem and the identity in Eq. (5.21), from Eq. (5.35), it can be easily derived that in the case of water of infinite depth, $\tilde{G}$ is given by

$$\begin{aligned}
\tilde{G}(x, y, x_0, y_0) &= -\sum_{n=0,I}^{II} \frac{i}{k_n \mathcal{C}_n}e^{ik_n|x-x_0|-k_n(y+y_0)} \\
&\quad - \frac{1}{\pi}\int_0^\infty \frac{M(\xi, y_0)M(\xi, y)e^{-\xi|x-x_0|}d\xi}{\xi\{\xi^2(1 - Q\xi^2 + D\xi^4)^2 + K^2\}},
\end{aligned} \qquad (5.39)$$

where $\mathcal{C}_n = F'_{pd}(k_n)/2K, F_{pd} \equiv k(1 - Nk^2 + Dk^4) - K = 0$. Further, it may be noted that k_n is the root of the dispersion relation $F_{pd} = 0$ as discussed in Chapter 3. Proceeding in a similar manner as in the case of infinite water depth, from Eq. (5.25) it can be derived that in the case of finite water depth, Green's function can be rewritten as

$$\begin{aligned}
\tilde{G} &= \ln\frac{r}{r'} - 2\int_0^\infty \frac{e^{-\xi h}}{\xi}\frac{\sinh\xi y\sinh\xi y_0}{\cosh\xi h}\cos\xi|x - x_0|d\xi \\
&\quad - 2\int_0^\infty \frac{\Omega_h\cosh\xi(h - y)\cosh\xi(h - y_0)}{\Omega_h\xi\sinh\xi h - K\cosh\xi h}\frac{\cos\xi|x - x_0|}{\cosh\xi h}d\xi,
\end{aligned} \qquad (5.40)$$

where Ω_h is the same as in Eq. (5.28). On simplification, and using the Cauchy reside theorem, an alternate form of the finite depth Green's function $\bar{G}$ is obtained as

$$\tilde{G}(x, y, x_0, y_0) = \sum_{n=0,I}^{II} A_n f_n(y)e^{ik_n|x-x_0|} + \sum_{n=1}^\infty B_n f_n(y)e^{-k_n|x-x_0|}, \qquad (5.41)$$

with $k_n = ik_n$ for $n = 1, 2, ...$ and k_n satisfying the dispersion relation $k(1 - Qk^2 + Dk^4) \tanh kh = K$ and A_n given by

$$A_n = \frac{-if_n(y_0)}{2k_nC_n}, \quad n = 0, I, II, \quad B_n = \frac{-f_n(y_0)}{2k_nC_n}, \quad n = 1, 2, ..., \tag{5.42}$$

In Eqs. (5.41) and (5.42), f_n and C_n are given by

$$f_n(y) = \begin{cases} \dfrac{\cosh k_n(h-y)}{\cosh k_n h}, & n = 0, I, II, \\ \dfrac{\cos k_n(h-y)}{\cos k_n h}, & n = 1, 2, , ..., \end{cases} \qquad C_n = \frac{F'_{hpd}(k_n)\tanh k_n h}{2K}.$$

It is easy to check that the time-harmonic Green's functions associated with flexural gravity wave motion given in Eqs. (5.39) and (5.41) in the cases of infinite and finite water depths are the same as discussed in Chapter 3 derived by the utilisation of the expansion formulae and the associated orthogonal mode-coupling relations.

5.3.2 Flexural gravity wavemaker problem

In the case of the flexural gravity wavemaker problem (as in the case of an unsteady gravity wavemaker problem in [84]) in the time domain, the velocity potential $\Phi(x, y, t)$ satisfies the governing Eq. (5.1) along with the surface boundary condition as in Eq. (5.7), the bottom boundary conditions as in Eqs. (5.4) and (5.5) and the initial conditions as in Eq. (5.6). The boundary condition on the vertical wavemaker is of the form

$$\frac{\partial \Phi}{\partial x} = U(y, t) \quad \text{on} \quad x = 0. \tag{5.43}$$

Thus, assuming that $\bar{\Phi}$ is the Laplace transform of $\Phi(x, y, t)$, it is easy to derive that $\bar{\Phi}$ satisfies the Laplace equation given by

$$\nabla^2 \bar{\Phi} = 0 \tag{5.44}$$

and the plate-covered boundary condition given by

$$D\frac{\partial^5 \bar{\Phi}}{\partial y^5} - Q\frac{\partial^3 \bar{\Phi}}{\partial y^3} + \frac{\partial \bar{\Phi}}{\partial y} - \frac{p^2}{g}\bar{\Phi} = 0, \quad \text{on} \quad y = 0, \quad 0 < x < \infty. \tag{5.45}$$

The condition on the wavemaker in Eq. (5.43) reduces to

$$\frac{\partial \bar{\Phi}}{\partial x} = \bar{U}(y, p) \quad \text{on} \quad x = 0. \tag{5.46}$$

Finally, $\bar{\Phi}$ satisfies the bottom boundary condition

$$\bar{\Phi}, |\nabla\bar{\Phi}| \to 0, \quad \text{as} \quad y \to \infty, \quad \text{in the case of infinite water depth} \tag{5.47}$$

and

$$\frac{\partial \bar{\Phi}}{\partial y} = 0 \quad \text{on} \quad y = h \quad \text{in the case of finite water depth.} \qquad (5.48)$$

Next, to determine $\bar{\Phi}$, Green's second identity will be used along with the Green's function derived in Subsection 5.3.1. Consider

$$\bar{G}^{mod}(x, y, x_0, y_0.p) = \bar{G}(x, y, x_0, y_0, p) + \bar{G}(x, y, -x_0, y_0, p), \qquad (5.49)$$

with zero normal velocity on the wavemaker, that is, $G_x^{mod}(0, y, x_0, y_0, p) = 0$, with G being Green's function defined in Eqs. (5.34) and (5.35) in the case of infinite water depth and finite water depth, respectively. Using Green's identity, the velocity potential $\bar{\Phi}(x, y, p)$ is obtained as

$$\bar{\Phi}(x_0, y_0, p) = -\int_0^\infty \bar{G}^{mod}(0, y, x_0, y_0, p)\bar{U}(y, p)dy + A(p) \qquad (5.50)$$

where

$$\begin{aligned}
A(p) &= \int_0^\infty \{\bar{G}^{mod}(x, 0, x_0, y_0, p)\bar{\Phi}_y(x, 0, p) \\
&\quad - \bar{\Phi}(x, 0, p)\bar{G}_y^{mod}(x, 0, x_0, y_0, p)\}dx,
\end{aligned} \qquad (5.51)$$

and the contribution from bottom boundary and at infinity become zero. Using the fact that $\bar{G}(x, y, x_0, y_0, p)$ is outgoing in nature at infinity and $\bar{G}^{mod}(x, y, x_0, y_0, p)$ satisfies

$$\begin{aligned}
\bar{G}_y^{mod}(0, 0, x_0, y_0) &= 2\bar{G}_y(0, 0, x_0, y_0), & \bar{G}_{xy}^{mod}(0, 0, x_0, y_0) &= 0, \\
\bar{G}_{yyy}^{mod}(0, 0, x_0, y_0) &= 2\bar{G}_{yyy}(0, 0, x_0, y_0), & \bar{G}_{xxxy}^{mod}(0, 0, x_0, y_0) &= 0,
\end{aligned}$$

it can be easily derived that

$$A(p) = -\frac{2}{p^2}\left[Q\bar{G}_y\Phi_{xy} - D\{\bar{G}_{yyy}\Phi_{xy} + \bar{G}_y\Phi_{xyyy}\}\right]_{(x,y)=(0,0)}. \qquad (5.52)$$

Using the fact that

$$\bar{G}^{mod}(0, y, x_0, y_0, p) = 2\bar{G}(0, y, x_0, y_0, p) = 2\bar{G}(x_0, y_0, 0, y; p),$$

from Eqs. (5.40) and (5.42), the velocity potential is obtained as

$$\bar{\Phi}(x_0, y_0, p) = -2\int_\Re \bar{G}(0, y, x_0, y_0; p)\bar{U}(y, p)dy + A(p), \qquad (5.53)$$

where $\Re$ is the semi-infinite interval $(0, \infty)$ in the case of infinite water depth and the finite interval $(0, h)$ in the case of finite water depth. The end behaviors are to be determined and depend on the specific problem. Once $\bar{\Phi}(x_0, y_0, p)$ is obtained using the convolution theorem and the inverse Laplace transform of $\bar{\Phi}(x_0, y_0, p)$, the function $\Phi(x_0, y_0, t)$ is obtained as

$$\Phi(x_0, y_0, t) = -2\int_\Re \int_0^t G(0, y, x_0, y_0, t - \tau)U(y; \tau)dyd\tau + a(t), \qquad (5.54)$$

where $a(t)$ is the Laplace inversion of $A(p)$. The details are deferred here.

5.4   Flexural gravity wave motion due to initial disturbances

In this section, the generation of flexural gravity waves by initial disturbances in the presence of axial force is formulated as an initial boundary value problem and analysed in two dimensions in the linearised water wave theory. Since the water surface is covered fully by a floating ice sheet/large flexible structure of uniform thickness, which is modeled as a thin elastic plate, the plate-covered surface boundary condition is of the fifth order. For unsteady motion associated with the generation of waves by initial disturbances, to handle the initial boundary value problem in the present section, both the Laplace and Fourier transforms are applied to reduce the problem to an ordinary differential equation in the transformed variables whose solutions are obtained in terms of complex oscillatory integrals. On the other hand, the Laplace transform and the Hankel transform are used to deal with the wave generation problem assuming that the motion is axi-symmetric in nature. Using the method of stationary phase, asymptotic results for large time and space are derived and analysed in particular cases to understand the far field wave amplitude and plate deflection.

5.4.1   Use of Laplace-Fourier transforms in wave motion

Here, the initial boundary value problem of generation of flexural gravity waves associated with a floating ice sheet in the presence of uniform compressive force N is analysed in two dimensions in the Cartesian coordinate system. The initial conditions are prescribed in the form of an initial depression or initial impulse on the elastic plate-covered surface at time $t = 0$. In the Cartesian coordinate system (x, y), the x-axis is taken as horizontal and the y-axis is vertically downward positive with the fluid occupying the region $-\infty < x < \infty, 0 < y < h$ in the case of water of finite depth h. The ice sheet of small thickness d is floating on the undisturbed water surface $-\infty < x < \infty$, $y = 0$. In the linearised theory of water waves, the fluid is assumed to be inviscid and incompressible and the motion is irrotational in nature. Thus, the associated velocity potential $\Phi(x, y, t)$ satisfies the PDE

$$\nabla^2 \Phi = 0, \quad \text{in the fluid domain,} \tag{5.55}$$

along with the linearised kinematic and the dynamic boundary conditions at the ice-covered mean free surface ($y = 0$ for $t > 0$) as given by

$$\frac{\partial \Phi}{\partial y} = \frac{\partial \eta}{\partial t}, \tag{5.56}$$

$$\frac{1}{g}\frac{\partial \Phi}{\partial t} = \left(D\frac{\partial^4}{\partial x^4} + Q\frac{\partial^2}{\partial x^2} + \gamma\frac{\partial^2}{\partial t^2} + 1 \right)\eta, \tag{5.57}$$

where the constants are the same as defined in Section 5.2. The no flow condition at the bottom boundary yields

$$\frac{\partial \Phi}{\partial y} = 0, \quad \text{on } y = h. \tag{5.58}$$

The initial conditions are

$$\Phi(x,0,0) = \Phi_0(x); \quad \eta(x,0) = \eta_0(x); \quad \eta'(x,0) = \eta_1(x), \tag{5.59}$$

where $\Phi_0(x)$, $\eta_0(x)$ and $\eta_1(x)$ are the initial impulse to the velocity potential, the initial surface elevation and the initial velocity of the ice sheet, respectively. Combining the kinematic and the dynamic upper surface boundary conditions (5.56) and (5.57), the linearised ice-covered surface boundary condition in the presence of uniform compressive force becomes

$$\left(D \frac{\partial^4}{\partial x^4} + Q \frac{\partial^2}{\partial x^2} + \gamma \frac{\partial^2}{\partial t^2} + 1 \right) \frac{\partial \Phi}{\partial y} = \frac{1}{g} \frac{\partial^2 \Phi}{\partial t^2}. \tag{5.60}$$

The initial boundary value problem will be solved by applying the Laplace and Fourier transforms and their inverse functions to obtain the velocity potential $\Phi(x,y,t)$ and thus the plate deflection $\eta(x,t)$. Before going into the details of the solution procedure, the combined Laplace-Fourier transform of a function $f(x,t)$ is defined by

$$\bar{\hat{f}}(k,p) = \int_{-\infty}^{\infty} \left\{ \int_0^{\infty} f(x,t)e^{-pt}dt \right\} e^{ikx}dx,$$

where p and k are the Laplace and Fourier transform variables, respectively, and the bar and hat refer to the Laplace and Fourier transforms, respectively. Applying the Laplace transform in time variable t and the Fourier transform in space variable x as defined above, from Eqs. (5.55)–(5.59), the initial value problem is reduced to a boundary value problem associated with an ordinary differential equation in $\bar{\hat{\Phi}}(k,y,p)$ given by

$$\frac{\partial^2 \bar{\hat{\Phi}}}{\partial y^2} - k^2 \bar{\hat{\Phi}} = 0, 0 < y < h, \tag{5.61}$$

$$\frac{\partial \bar{\hat{\Phi}}}{\partial y} = 0, \quad \text{on } y = h, \tag{5.62}$$

$$\frac{\partial \bar{\hat{\Phi}}}{\partial y} + \hat{\eta}_0 = p\bar{\hat{\eta}} \quad \text{on } y = 0, \tag{5.63}$$

and

$$\hat{\Phi}_0 + g(Dk^4 - Qk^2 + 1)\bar{\hat{\eta}} = p\bar{\hat{\Phi}}, \quad \text{on } y = 0. \tag{5.64}$$

With the solution of the differential equation in Eq. (5.61) satisfying the conditions in Eqs. (5.62)–(5.64), the transformed function $\hat{\bar{\Phi}}(k, y, p)$ is obtained as

$$\hat{\bar{\Phi}}(k, y; p) = \frac{\hat{\bar{\eta}}(Dk^4 - Qk^2 + 1)g + \hat{\Phi}_0(k)\cosh k(h - y)}{p} \cdot \frac{\cosh k(h-y)}{\cosh kh}, \qquad (5.65)$$

where $\hat{\bar{\Phi}}(k, y, p)$ represents the Laplace-Fourier transform of $\Phi(x, y, t)$ and $\eta(x, t)$, respectively. Here, also, the terms containing γ are neglected as discussed earlier. Substituting for $\hat{\bar{\Phi}}(k, y; p)$ from Eq. (5.65) into Eq. (5.63), $\hat{\bar{\eta}}$ is obtained as

$$\hat{\bar{\eta}} = \frac{\hat{\eta}_0(k)}{2}\left\{\frac{1}{p - i\alpha_1} + \frac{1}{p - i\alpha_2}\right\} + \frac{i\hat{\Phi}_0(k)k \tanh kh}{2}\left\{\frac{1}{p - i\alpha_1} - \frac{1}{p - i\alpha_2}\right\}, \qquad (5.66)$$

where

$$\alpha_{1,2} = \pm\sqrt{(Dk^4 - Qk^2 + 1)gk \tanh kh}, \qquad (5.67)$$

with the subscripts 1 and 2 in $\alpha_{1,2}$ corresponding to the $+$ and $-$ signs in the right-hand side, respectively, hereafter.

Next, applying the inverse Laplace transform and then the inverse Fourier transform to Eq. (5.66), the surface elevation η is obtained as

$$\begin{aligned}
\eta(x, t) &= \frac{1}{4\pi}\int_{-\infty}^{\infty}\left[\hat{\eta}_0(k)(e^{i\alpha_1 t} + e^{i\alpha_2 t})\right. \\
&\quad + \left. i\hat{\Phi}_0(k)\left(\frac{k \tanh kh}{g(Dk^4 - Qk^2 + 1)}\right)^{1/2}(e^{i\alpha_1 t} - e^{i\alpha_2 t})\right]e^{ikx}dk. \quad (5.68)
\end{aligned}$$

To evaluate the integral in Eq. (5.68) for η, the specific initial condition has to be prescribed. Assume that the initial depression is concentrated at the origin, which gives $\eta_0(x) = h^2\delta(x)$; $\Phi_0(x) = 0$ where $\delta(x)$ is the Dirac-delta function. In this case, the flexural gravity wave elevation $\eta(x, t)$ is obtained as

$$\eta(x, t) = \frac{h^2}{4\pi}\int_{-\infty}^{\infty}\left\{e^{it(\alpha_1 + kx/t)} + e^{it(\alpha_2 + kx/t)}\right\}dk. \qquad (5.69)$$

Next, assume that the initial impulse is concentrated at the origin, $\eta_0(x) = 0, \Phi_0(x) = h^2 P\delta(x)$, where P is the total impulse. In this case, flexural gravity wave elevation $\eta(x, t)$ is given by

$$\eta(x, t) = -\frac{Ph^2}{4\pi i}\int_{-\infty}^{\infty}\sqrt{\frac{k \tanh kh}{g(Dk^4 - Qk^2 + 1)}}\left\{e^{it(\alpha_1 + kx/t)} - e^{it(\alpha_2 + kx/t)}\right\}dk. \qquad (5.70)$$

For the sake of convenience, various nondimensional parameters are introduced as $\tilde{x} = x/h$, $\tilde{t} = (g/h)^{1/2}t$, $\tilde{\eta}(\tilde{x}, \tilde{t}) = \eta(x, t)/h$, $\tilde{D} = D/h^4$, $\tilde{Q} = Q/h^2$ and $\tilde{P} = P/(gh)^{1/2}$. Thus, dropping the $\tilde{\ }$, the simplified version of the expressions in Eqs. (5.69) and (5.70) are given by

$$\eta(x,t) = \frac{h}{4\pi} \int_{-\infty}^{\infty} \left\{ e^{itf_1(kh)} + e^{itf_2(kh)} \right\} dk, \tag{5.71}$$

$$\eta(x,t) = -\frac{Ph}{4\pi i} \int_{-\infty}^{\infty} \sqrt{\frac{kh\tanh kh}{Dk^4h^4 - Qk^2h^2 + 1}} \left\{ e^{itf_1(kh)} - e^{itf_2(kh)} \right\} dk, \tag{5.72}$$

with

$$f_{1,2}(kh) = \pm\sqrt{(Dk^4h^4 - Qk^2h^2 + 1)kh\tanh kh} + \frac{x}{t}kh. \tag{5.73}$$

It may be noted that integrands associated with the above integrals are oscillatory in nature and the asymptotic forms of the integrals are more suitable for certain physical applications. Next, the integrals will be evaluated asymptotically for large distance and time x and t, respectively, by the application of the method of stationary phase to understand the asymptotic behavior of the surface profile $\eta(x,t)$ as in [84]. Thus, in the integrals as in Eqs. (5.71) and (5.72), the stationary points will come from the roots of the equations

$$f_1'(kh) = 0 \quad \text{and} \quad f_2'(kh) = 0, \tag{5.74}$$

where

$$f_{1,2}'(kh) = \pm\left\{ \frac{1 - 3Qk^2h^2 + 5Dk^4h^4}{2(1 - Qk^2h^2 + Dk^4h^4)^{1/2}} \left(\frac{\tanh kh}{kh} \right)^{1/2} \right.$$
$$\left. + \frac{(1 - Qk^2h^2 + Dk^4h^4)^{1/2}}{2} \left(\frac{kh}{\tanh kh} \right)^{1/2} \frac{1}{\cosh^2 kh} \right\} + \frac{x}{t}. \tag{5.75}$$

From numerical computations it is observed that $f_1'(kh) = 0$ has a single real root α_1 (say) and $f_2'(kh) = 0$ has two real roots α_2 and α_3 (say). Hence, using the relations $f_1'(-kh) = -f_2'(kh)$ and $f_2'(-kh) = -f_1'(kh)$, and then applying the method of stationary phase to the integrals in Eqs. (5.71) and (5.72), the surface elevations $\eta(x,t)$ for large space and time are derived as

$$\eta(x,t) = \sum_{i=1}^{3} \frac{1}{2\sqrt{2\pi}} \left(\frac{1}{tf_i''(\alpha_i)} \right)^{1/2} \cos\left\{ tf_i(\alpha_i) + \frac{\pi}{4} \right\} \tag{5.76}$$

and

$$\eta(x,t) = \sum_{i=0}^{3} -\frac{P}{2\sqrt{2\pi}} \left(\frac{1}{tf_i''(\alpha_i)} \right)^{1/2} \left(\frac{\alpha_i \tanh \alpha_i}{D\alpha_i^4 - Q\alpha_i^2 + 1} \right)^{1/2} \sin\left\{ tf_i(\alpha_i) + \frac{\pi}{4} \right\}, \tag{5.77}$$

respectively, with $f_3(\alpha_3)$ corresponding to $f_2(\alpha_3)$. Eq. (5.76) gives the asymptotic form of the nondimensional surface elevation where the initial depression is concentrated at the origin, while Eq. (5.77) gives the surface elevation where the initial impulse is concentrated at the origin for large values of space and time x and t, respectively. The solution of the wave generation by initial disturbances in the case of infinite water depth can be obtained from the finite depth results by considering $h \to \infty$.

5.4.2 Use of the Laplace-Hankel transform in wave motion

In the present study, the initial boundary value problem associated with the generation of flexural gravity waves in the presence of uniform compressive force is discussed in the cylindrical polar coordinate assuming the motion is axi-symmetric. Here, the flexural gravity wave motion is generated due to the impingement of surface wave with a floating elastic plate in finite water depth. As with the analysis of the previous subsection, here the initial conditions are prescribed in the form of initial depression or initial impulse on the elastic plate-covered surface at time $t = 0$. In the cylindrical coordinate system (r, θ, y) r is along the radial direction, θ is along the transverse axis and y-axis is vertically downward positive with the fluid occupying the region $0 < r < \infty$, $0 < y < h$ in the case of water of finite depth h. The ice sheet of small thickness d is floating on the undisturbed water surface $0 < r < \infty, 0 < \theta < 2\pi$ and $y = 0$. Thus, under the assumption of the linearised theory of water waves and assuming that the motion is axi-symmetric in nature, the velocity potential $\Phi(r, y, t)$ satisfies the PDE

$$\frac{1}{r}\frac{\partial}{\partial r}\left(r\frac{\partial\Phi}{\partial r}\right) + \frac{\partial^2\Phi}{\partial y^2} = 0, \quad 0 < y < h, \ r > 0. \tag{5.78}$$

The assumption of no gap between the floating elastic plate and the water surface at any time yields the kinematic condition given by

$$\frac{\partial\Phi}{\partial y} = \frac{\partial\eta}{\partial t}, \quad \text{on} \quad y = 0. \tag{5.79}$$

Further, the linearised condition on the floating elastic plate-covered mean free surface ($y = 0$ for $t > 0$) is given by

$$\left(D\nabla_r^4 + Q\nabla_r^2 + 1\right)\frac{\partial\Phi}{\partial y} = \frac{1}{g}\frac{\partial^2\Phi}{\partial t^2}, \quad \text{on} \quad y = 0, \ 0 < y < h, \tag{5.80}$$

where

$$\nabla_r^2 = \frac{1}{r}\frac{\partial}{\partial r}\left(r\frac{\partial}{\partial r}\right)$$

assuming that the inertial term is neglected (as discussed in the previous subsection), with D and Q being the same as in Eq. (5.3). The no flow condition at the bottom boundary yields

$$\frac{\partial\Phi}{\partial y} = 0, \quad \text{on} \quad y = h. \tag{5.81}$$

The initial conditions are of the forms

$$\Phi(r,0,0) = -\frac{G_0(r)}{\rho} \quad \text{and} \quad \Phi_t(r,0,0) = (1 + Q\nabla_r^2 + D\nabla_r^4)gG_1(r), \quad (5.82)$$

where $G_0(r)$ and $G_1(r)$ are the initial impulse to the velocity potential and the initial depression of the ice sheet, respectively. The form of $\Phi_t(r,0,0)$ in Eq. (5.82) is chosen for a simplified form of the expression for the plate deflection $\eta(x,t)$, unlike the usual form of the initial data given in the previous subsection in Eq. (5.59).

Further, it may be noted that in a realistic physical problem, $G_0(r)$ and $G_1(r)$ need not be nonzero at the same time $t = 0$. Further, it may be noted that in the case of the initial value problem with an initial axial symmetric depression, $G_0(r) = 0$ and $G_1(r) \neq 0$. On the other hand, in the case of the axially symmetric impulse, $G_0(r) \neq 0$ and $G_1(r) = 0$. Using the dimensionless variables, $r' = r/l, D' = D/l^4, Q' = Q/l^2, y' = y/l, t' = t\sqrt{g/l}, \Phi' = \Phi/(l\sqrt{gl}), \eta' = \eta/l$, where l is the characteristic length, and g is the acceleration due to gravity and dropping the primes as in [86], for the initial value problem in terms of the nondimensional Φ, it is easy to see that Φ satisfies Eq. (5.78) along with the boundary conditions in Eqs. (5.79) and (5.81). However, the surface condition in Eq. (5.80) is modified as

$$\left(D\nabla_r^4 + Q\nabla_r^2 + 1\right)\frac{\partial\Phi}{\partial y} = \frac{\partial^2\Phi}{\partial t^2}, \quad \text{on} \quad y = 0, \ 0 < y < h, \quad (5.83)$$

while the initial conditions in Eq. (5.82) are modified as

$$\Phi(r,0,0) = -\frac{G_0(r)}{\rho} \text{ and } \Phi_t(r,0,0) = (1 + Q\nabla_r^2 + D\nabla^4)G_1(r). \quad (5.84)$$

The Laplace transform of $\Phi(r,y,t)$ denoted as $\bar{\Phi}(r,y,p)$ is defined as

$$\bar{\Phi}(r,y,p) = \int_0^\infty \Phi(x,y,t)e^{-pt}dt, \quad \text{Re}\{p\} > 0, \quad (5.85)$$

where p is the Laplace transform variable and the bar refers to the Laplace transform. Applying the Laplace transform in time variable t as defined above, the initial value problem in terms of the nondimensional velocity potential $\Phi(r,y,t)$ is reduced to the boundary value problem in $\bar{\Phi}(r,y,p)$ which satisfies the partial differential equation

$$\frac{1}{r}\frac{\partial}{\partial r}\left(r\frac{\partial\bar{\Phi}}{\partial r}\right) + \frac{\partial^2\bar{\Phi}}{\partial y^2} = 0, \quad 0 < y < h, \ r > 0, \quad (5.86)$$

along with the boundary condition at the mean free surface being given by

$$\{D\nabla_r^4 + Q\nabla_r^2 + 1\}\frac{\partial\bar{\Phi}}{\partial y} = p^2\bar{\Phi} - \{D\nabla_r^4 + Q\nabla_r^2 + 1\}G_1(r) - \frac{pG_0(r)}{\rho}, \quad \text{on } y = 0$$

$$(5.87)$$

and the bottom boundary condition being given by

$$\frac{\partial \bar{\Phi}}{\partial y} = 0, \quad \text{on } y = h. \tag{5.88}$$

The Laplace transform of the kinematic condition in Eq. (5.79) yields

$$\frac{\partial \bar{\Phi}}{\partial y} = p\bar{\eta}(x,p) - \eta(x,0), \quad \text{on } y = 0. \tag{5.89}$$

The boundary value problem in $\bar{\Phi}$ satisfying Eq. (5.86) along with the boundary conditions in Eqs. (5.87)–(5.89) is handled for solution with the help of the Hankel transform. The Hankel transform of the function $\bar{\Phi}(r, y, p)$ denoted as $\Psi(k, y, p)$ is defined as

$$\Psi(k, y, p) = \int_0^\infty r\bar{\Phi}(r, y, p) J_0(kr)dr, \quad k > 0. \tag{5.90}$$

The Hankel transform of Eq. (5.86) along with the conditions in Eqs. (5.87) and (5.88) yields

$$\frac{d^2\Psi}{dy^2} - k^2\Psi = 0, 0 < y < h \tag{5.91}$$

subject to the boundary conditions

$$p^2\Psi - \{DL^2(k) + QL(k) + 1\}\frac{\partial\Psi}{\partial y}$$

$$= \{DL^2(k) + QL(k) + 1\}\hat{G}_1(k) - \frac{p}{\rho}\hat{G}_0(k), \quad \text{on } y = 0, \tag{5.92}$$

where $L(k) = d^2/dk^2 - k^2$, $\hat{G}_j(k)$ is the Hankel transform of $G_j(r)$ for $j = 1, 2$. Further,

$$\frac{\partial\Psi}{\partial y} = 0, \quad \text{on } y = h. \tag{5.93}$$

Further, the Hankel transform of Eq. (5.89) yields

$$\frac{\partial\Psi}{\partial y} = p\hat{\bar{\eta}}(k, p) - \hat{\eta}(k, 0), \quad \text{on } y = 0. \tag{5.94}$$

Thus, solving Eq. (5.91) subject to the boundary conditions in Eqs. (5.92) and (5.93), it can be derived that

$$\Psi(k, y; p) = \frac{(Dk^4 - Qk^2 + 1)\hat{G}_1(k) - (p/\rho)\hat{G}_0(k)}{p^2 + c^2} \frac{\cosh k(h - y)}{\cosh kh}, \tag{5.95}$$

where $c = \sqrt{(Dk^4 - Qk^2 + 1)k \tanh kh}$. Eqs. (5.94) and (5.95) yield

$$\hat{\bar{\eta}}(k, p) = \frac{1}{p}\left\{\frac{(p/\rho)\hat{G}_0(k) - (Dk^4 - Qk^2 + 1)\hat{G}_1(k)}{p^2 + c^2}\right\}k \tanh kh - \hat{\eta}(k, 0).$$

$$\tag{5.96}$$

Next, the two specific examples will be discussed separately. Assuming that the displacement is concentrated at the origin, that is, $\hat{G}_0 = 0$ and $\hat{G}_1(k) = -\hat{\eta}(k,0) = 1/2\pi$, Eq. (5.96) yields

$$\hat{\bar{\eta}}(k,p) = \frac{1}{2\pi}\frac{p}{p^2 + c^2}, \qquad (5.97)$$

whose inverse Laplace transform yields

$$\hat{\eta}(k,t) = \frac{1}{2\pi}\cos ct. \qquad (5.98)$$

Then, the inverse Hankel transform of Eq. (5.98) yields

$$\eta(r,t) = \frac{1}{2\pi}\int_0^\infty k\cos ct\, J_0(kr)dk. \qquad (5.99)$$

Next, assuming $\hat{G}_0 = 1/2\pi$ and $\hat{G}_1 = \hat{\eta}(k,0) = 0$, Eq. (5.96) yields

$$\hat{\bar{\eta}}(k,p) = \frac{1}{2\pi\rho}\frac{k\tanh kh}{c}\frac{c}{p^2 + c^2}, \qquad (5.100)$$

whose Laplace inversion yields

$$\hat{\eta}(k,t) = \frac{k^{1/2}}{2\pi p}\left(\frac{\tanh kh}{Dk^4 - Qk^2 + 1}\right)^{1/2}\sin ct. \qquad (5.101)$$

Then, the inverse Hankel transform of Eq. (5.101) yields

$$\eta(r,t) = \frac{1}{2\pi\rho}\int_0^\infty k^{3/2}\left(\frac{\tanh kh}{Dk^4 - Qk^2 + 1}\right)^{1/2}\sin ct\, J_0(kr)dk. \qquad (5.102)$$

As with the case of wave generation in Cartesian coordinates as discussed in the previous subsection, here the asymptotic forms of $\eta(r,t)$ in Eqs. (5.99) and (5.102) are obtained for large distance r and time t but finite r/t by the application of the method of stationary phase (as in [24]). Using the identity

$$J_0(kr) = \frac{2}{\pi}\int_0^{\pi/2}\cos(kr\cos\theta)d\theta, \qquad (5.103)$$

the plate deflection $\eta(r,t)$ in Eq. (5.99) is rewritten as

$$\eta(r,t) = \frac{1}{2\pi}\int_0^\infty\int_0^{\pi/2}k\left(e^{ict} + e^{-ict}\right)\left(e^{ikr\cos\theta} + e^{-ikr\cos\theta}\right)d\theta dk. \qquad (5.104)$$

Application of the method of stationary phase to the θ-integral in Eq. (5.104), yields

$$\eta(r,t) \approx \frac{1}{4\pi^2}\int_0^\infty\left(\frac{2\pi k}{r}\right)^{1/2}\left(e^{ikr + i\pi/4} + e^{-ikr - i\pi/4}\right)\left(e^{ict} + e^{-ict}\right)dk. \qquad (5.105)$$

Writing $f_1(k) = c + k\dfrac{r}{t} + \dfrac{\pi}{4t}$ and $f_2(k) = -c + k\dfrac{r}{t} + \dfrac{\pi}{4t}$, it is easy to see that $f_1'(k)$ has one real root α_1 (say) and $f_2'(k)$ has two real roots α_2 and α_3 (say). Thus, by applying the method of stationary phase to the $k-$ integral in Eq. (5.105), it is found that

$$\eta(r,t) \approx \frac{1}{\pi} \sum_{i=1}^{3} \left(\frac{\alpha_i}{r}\right)^{1/2} \left(\frac{1}{t|f_1''(\alpha_i)|}\right)^{1/2} \cos\{tf_i(\alpha_i) + \pi/4\}. \qquad (5.106)$$

Next, the method of stationary phase is applied to the integral in Eq. (5.102). Proceeding in a similar manner and using the integral form for $J_0(kr)$ as in Eq. (5.103) and applying the method of stationary phase to the $\theta-$ integral, it can be easily derived that

$$\eta(r,t) \approx \left(\frac{2\pi}{r}\right)^{1/2}\frac{1}{\rho}\int_0^\infty ka(k)\left(e^{ikr+i\pi/4} + e^{-ikr-i\pi/4}\right)\left(\frac{e^{ict} - e^{-ict}}{2i}\right)dk, \qquad (5.107)$$

where $f_1(k)$ and $f_2(k)$ are the same as in Eq. (5.106) and $a(k) = \sqrt{k\tanh kh}/(Dk^4 - Qk^2 + 1)$. As with the previous case, in this case also, it can be checked numerically that $f_1'(k) = 0$ has one real root and $f_2'(k) = 0$ has two real roots. Thus, by the method of stationary phase, it can be easily derived that

$$\eta(r,t) \approx \frac{-1}{\rho\pi}\frac{1}{r^{1/2}} \sum_{i=1}^{3} \alpha_i \left(\frac{\tanh \alpha_i h}{1 - Q\alpha_i^2 + D\alpha_i^4}\right)^{1/2}$$

$$\times \left(\frac{1}{t|f_1''(\alpha_i)|}\right)^{1/2} \sin\{tf_1(\alpha_i) + \pi/4\}. \qquad (5.108)$$

It may be noted that in Eqs. (5.106) and (5.108), $f_3(\alpha_3)$ corresponds to $f_2(\alpha_3)$. Further, it is assumed that in Eq. (5.108), $1 - Q\alpha_i^2 + D\alpha_i^4 \neq 0$ for $i = 1, 2, 3$ which will be true for all realistic physical situations. Further, the solution associated with the problem of wave generation by initial disturbances in the case of infinite water depth can be obtained from the finite depth results by considering $h \to \infty$.

5.5 Conclusion

In the present chapter, solution techniques for a class of time-dependent initial value problems are discussed based on the suitable application of integral transform along with the method of stationary phase. The time domain Green's function is obtained for homogeneous fluid of constant density in the two-dimensional Cartesian coordinate system with the help of the Laplace transform. With the application of the Laplace transform, a class of initial boundary value problems is reduced to boundary problems in the transformed variable whose solution is obtained with the help of Green's function technique. Further, using the time domain Green's function, often initial boundary value problems are solved with the help of Green's identity. The wave generation problems are handled for solution with the help of Laplace-Fourier transforms in two dimensions in the Cartesian coordinate system, and combined Laplace and Hankel transforms are used to deal with axi-symmetric problems. The convolution theorem associated with the Laplace transform has played a significant role while considering the inverse Laplace transforms. Further, in many situations, asymptotic solutions are obtained with the help of the method of stationary phase. The discussed approach can be generalised to deal with wave structure interaction problems in three dimensions, some of which are left as exercises at the end of the chapter. Another interesting class of problems in time domain analysis is on moving loads on the floating elastic plates/ice sheets. The application of moving loads on floating ice sheets with historical perspective is discussed in [127]. Some of the recent works on moving loads on floating ice sheets can be found in [97] and [111] and the cited literature. Often the time domain mode-expansion method is used to deal with unsteady wave structure interaction problems (see [7], [65], [105] and [110]). In the present chapter, basic techniques for a class of wave structure interaction problems are discussed based on small amplitude wave theory. However, there is a significant interest to deal with time domain wave structure interaction problems based on shallow water wave theory and finite amplitude wave theory, the details of which are deferred in the present monograph (see for details [45], [134] and [135]). Various physical problems of fluid structure interactions in time domain can be handled for solution generalising the methods demonstrated and the other methods which are briefly discussed in the chapter.

5.6   Examples and exercises

Exercise 5.1 Derive the time domain Green's function to deal with wave interaction with floating ice sheet in cylindrical polar coordinate assuming the motion is axi-symmetric in the cases of water of both finite and infinite depths.

Exercise 5.2 Derive the time-domain Green's function to deal with wave interaction with floating ice sheet in the three-dimensional Cartesian coordinate system in the cases of water of both finite and infinite depths.

Exercise 5.3 Derive the time domain Green's function to deal with oblique wave interaction with floating ice sheet in the Cartesian coordinate system in the cases of water of both finite and infinite depths.

Exercise 5.4 Derive the time domain Green's function to deal with wave interaction with floating ice sheet in a two-layer fluid having an interface in two- and/or three-dimensional Cartesian coordinate system in the cases of water of both finite and infinite depths.

Exercise 5.5 Derive the time domain Green's function to deal with oblique wave interaction with floating ice sheet in a two-layer fluid having an interface in Cartesian coordinate system in the cases of water of both finite and infinite depths.

Exercise 5.6 Derive the time domain Green's function to deal with gravity wave interaction with a flexible submerged horizontal plate in two- and/or three-dimensional Cartesian coordinate system in the cases of water of both finite and infinite depths.

Exercise 5.7 Discuss the generation of flexural gravity waves in three-dimensions due to an initial depression/impuse in the three-dimensional Cartesian coordinate system in the cases of water of both finite and infinite depths.

Exercise 5.8 Discuss the generation of flexural gravity waves in three dimensions due to an initial depression/impuse in the three-dimensional Cartesian coordinate system in the cases of water of both finite and infinite depths in a two-layer fluid having an interface and a floating elastic plate-covered free surface.

Exercise 5.9 Discuss the generation of flexural gravity waves in two/three dimensions in the presence of uniform current in the cases of both finite and infinite water depths in the two/three dimensional-Cartesian coordinate system.

Exercise 5.10 Discuss the generation of flexural gravity waves in two-/three-dimensional Cartesian coordinate system in the cases of both finite and infinite water depths due to a steadily moving load.

6

Shallow water approximation

6.1 General introduction

In the previous chapters, all the problems have been studied under the assumption of small amplitude water wave theory and small amplitude structural response. However, in the present chapter, gravity wave interactions with large flexible floating structures are analysed under the assumption of small amplitude shallow water approximation and structural response. The basic equations of wave structure interaction problems under shallow water approximation are briefly derived. Then, the solution procedures associated with the interaction of oblique wave scattering by floating elastic plates are discussed in several cases by matching the velocity and pressure at the interfaces. The utility of wide spacing approximation to deal with wave scattering by multiple floating structures is demonstrated in several cases. Although, the main emphasis in this chapter is on problems of single frequency, time-dependent problems associated with the hydroelastic analysis of large floating structures under shallow water approximations are briefly reviewed. Further, various methods associated with the reduction of the structural response of large floating structures is briefly reviewed.

6.1.1 Basic equations of linear long wave theory

Unlike in the case of small amplitude surface gravity wave problems in which the wave length is small compared to the depth of water, in the special case when the wave length is large compared to the water depth, the waves are known as long waves. A brief description of the linearised long wave equations is derived here under shallow water approximation. In the case of long waves, the horizontal component of the fluid velocity is much larger compared to the vertical component. In this case, for a tank of finite water depth $h(x, z)$, which is infinitely large in the x and z directions, the y-axis is along the water depth which is considered positive in the downward direction. Assuming that (u, v, w) are the components of velocity $\vec{q}$, the equation of motion in the case of linearised long waves takes the form (as in [28], [129])

$$\frac{\partial u}{\partial t} + g\frac{\partial \zeta}{\partial x} = 0, \quad \frac{\partial w}{\partial t} + g\frac{\partial \zeta}{\partial z} = 0 \tag{6.1}$$

and the linearised hydrodynamic pressure yields

$$p = p_0 + \rho g(\zeta - y), \tag{6.2}$$

with $p = p_{atm}$ at the free surface $y = \zeta(x, z, t)$. In Eq. (6.2), p_{atm} is the constant atmospheric pressure, ζ is the surface elevation and g is the acceleration due to gravity. The equation of continuity for incompressible fluid yields

$$\frac{\partial u}{\partial x} + \frac{\partial v}{\partial y} + \frac{\partial w}{\partial z} = 0. \tag{6.3}$$

As discussed in Chapter 1, on the free surface $y = \zeta(x, z, t)$ and the kinematic condition yields

$$\frac{\partial \zeta}{\partial t} + u\frac{\partial \zeta}{\partial x} + w\frac{\partial \zeta}{\partial z} = v(x, y, t). \tag{6.4}$$

Integrating the continuity equation in Eq. (6.3) over water depth and using the kinematic condition in Eq. (6.4), the continuity equation associated with the linearised long wave equation in two dimensions is obtained as

$$\frac{\partial \zeta}{\partial t} - \frac{\partial(hu)}{\partial x} - \frac{\partial(hw)}{\partial z} = 0, \tag{6.5}$$

where $h(x, z)$ is the water depth and $y = h(x, z)$ is the bottom bed of the channel. From Eqs. (6.1) and (6.5), the two-dimensional long wave equation having a free surface in the case of uniform depth h (h is considered a constant) is obtained as

$$\frac{\partial^2 \zeta}{\partial x^2} + \frac{\partial^2 \zeta}{\partial z^2} = \frac{1}{c^2}\frac{\partial^2 \zeta}{\partial t^2}, \tag{6.6}$$

where $c^2 = gh$, with c being the speed of propagation in the case of linearised long waves. In the case of potential flow

$$u = \Phi_x, w = \Phi_z. \tag{6.7}$$

Hence, in the case of irrotational motion, from Eqs. (6.5) and (6.7), the continuity equation for the long wave in uniform water depth is obtained as

$$\frac{\partial \zeta}{\partial t} - h\left\{\frac{\partial^2 \Phi}{\partial x^2} + \frac{\partial^2 \Phi}{\partial z^2}\right\} = 0. \tag{6.8}$$

From Eq. (6.1), the long wave equation of motion in terms of the velocity potential is written as

$$\frac{\partial \Phi}{\partial t} - g\zeta = 0. \tag{6.9}$$

Eliminating $\zeta(x, t)$, from Eq. (6.8) and (6.9), the two-dimensional linearised long wave equation in terms of the velocity potential Φ in uniform water depth h is derived as

$$\frac{1}{c^2}\frac{\partial^2 \Phi}{\partial t^2} - \left\{\frac{\partial^2 \Phi}{\partial x^2} + \frac{\partial^2 \Phi}{\partial z^2}\right\} = 0. \tag{6.10}$$

For simplicity, it is assumed that the wave motion is simple harmonic in time with angular frequency w and that the oblique incident wave makes an angle θ with the positive direction of x-axis. Hence, the velocity potential $\Phi(x, y, z, t)$ and the free surface elevation are written as $\Phi(x, y, z, t) = Re[\phi(x, y)e^{ik_z z - iwt}]$ and $\zeta(x, z, t) = Re[\eta(x)e^{ik_z z - iwt}]$. Thus, the equation of motion in Eq. (6.6) yields

$$\left\{ \mathcal{L}(\partial_x) + \frac{w^2}{gh} \right\} \eta = 0, \qquad (6.11)$$

where

$$\mathcal{L}(\partial_x) \equiv \frac{\partial^2}{\partial x^2} - k_z^2.$$

Further, Eq. (6.10) yields

$$\left\{ \mathcal{L}(\partial_x) + \frac{w^2}{gh} \right\} \phi = 0. \qquad (6.12)$$

Assuming that the progressive wave solution for ϕ is of the form $\phi(x, y) = \psi(y)e^{-ik_x x}$, it can be easily derived that

$$k^2 = \frac{w^2}{gh}, \qquad (6.13)$$

where $k_x^2 + k_z^2 = k_0^2$ with $k_x = k_0 \cos\theta$, $k_z = k_0 \sin\theta$ and $\pm k_0$ are the real roots of Eq. (6.13) in k. Eq. (6.13) is the dispersion relation associated with the gravity wave under shallow water approximation, often referred to as the linearised long wave dispersion relation. Thus, the phase and group velocities associated with the linearised long wave in the open water region denoted as c and c_g are given by

$$c = \sqrt{gh}, \quad c_g = c. \qquad (6.14)$$

Various applications of linear long wave theory are demonstrated in [28] and [29]. However, the equation of continuity for the long wave as in Eq. (6.5) will be used in the derivation of the equation of motion associated with wave structure interaction problems, the details of which will be discussed in the next subsection.

6.1.2 Basic equations of wave structure interaction problems

In Chapter 1, basic equations and boundary conditions associated with wave structure interaction have been discussed under the assumption of small amplitude water wave theory and structural response. In the present subsection, the basic equations associated with wave structure interaction problems under shallow water approximation are discussed in brief.

In the case of a homogeneous fluid having an elastic plate-covered free surface, the linearised Bernoulli's equation of motion yields

$$P_H = -\rho \frac{\partial \Phi}{\partial t} + \rho g \zeta. \tag{6.15}$$

Further, as discussed in Chapter 1, in the presence of compressive force, the elastic plate equation yields

$$\left(EI\nabla^4_{xz} + N\nabla^2_{xz} + \rho_i d \frac{\partial^2}{\partial t^2} \right)\zeta = P_s, \tag{6.16}$$

where P_s is the differential pressure acting on the floating elastic plate and $\zeta(x, z, t)$ is the deflection of the elastic plate. Assuming that on the plate-covered mean free surface, the atmospheric pressure is constant, Eqs. (6.15) and (6.16) yield the linearised long wave equation of motion in the plate-covered region, which is given by

$$\left\{ D\nabla^4_{xz} + Q\nabla^2_{xz} + \left(1 + \frac{\rho_i d}{\rho g} \frac{\partial^2}{\partial t^2} \right) \right\} \zeta = \frac{1}{g} \frac{\partial \Phi}{\partial t}, \tag{6.17}$$

where $D = EI/\rho g$ and $Q = N/\rho g$. Further, Φ satisfies the continuity equation associated with the long wave given in Eq. (6.8). Thus, eliminating $\Phi(x, y, z, t)$ from Eqs. (6.8) and (6.17), the long wave equation in the plate-covered region is obtained as

$$\left\{ D\nabla^6_{xz} + Q\nabla^4_{xz} + \left(1 + \frac{\rho_i d}{\rho g} \frac{\partial^2}{\partial t^2} \right)\nabla^2_{xz} \right\} \zeta = \frac{1}{gh} \frac{\partial^2 \zeta}{\partial t^2}. \tag{6.18}$$

Next, eliminating ζ from the continuity equation in Eq. (6.8) and the equation of motion in Eq. (6.17), the long wave equation on the plate-covered region in terms of the velocity potential is written as

$$\left\{ D\nabla^6_{xz} + Q\nabla^4_{xz} + \left(1 + \frac{\rho_i d}{\rho g} \frac{\partial^2}{\partial t^2} \right)\nabla^2_{xz} \right\} \Phi = \frac{1}{gh} \frac{\partial^2 \Phi}{\partial t^2}. \tag{6.19}$$

For simplicity, it is assumed that the wave motion is simple harmonic in time with angular frequency ω and that the oblique incident wave makes an angle θ with the positive direction of the x-axis. Hence, the velocity potential $\Phi(x, y, z, t)$ and the deflection of the plate-covered surface are written as $\Phi(x, y, z, t) = Re[\phi(x, y)e^{ip_z z - i\omega t}]$ and $\zeta(x, z, t) = Re[\eta(x)e^{ip_z z - i\omega t}]$. Thus, the equation of motion in Eq. (6.18) yields

$$\left\{ D\mathcal{M}^3(\partial_x) + Q\mathcal{M}^2(\partial_x) + \left(1 - \frac{\rho_i d\omega^2}{\rho g} \right)\mathcal{M}(\partial_x) + \frac{\omega^2}{gh} \right\}\eta = 0, \tag{6.20}$$

where

$$\mathcal{M}(\partial_x) = \frac{\partial^2}{\partial x^2} - p_z^2.$$

Proceeding in a similar manner, Eq. (6.19) yields

$$\left\{ D\mathcal{M}^3(\partial_x) + Q\mathcal{M}^2(\partial_x) + \left(1 - \frac{\rho_i d\omega^2}{gh}\right)\mathcal{M}(\partial_x) + \frac{\omega^2}{gh}\right\}\phi = 0. \qquad (6.21)$$

Assuming that the progressive wave solution for ϕ is of the form $\phi(x, y) = \psi(y)e^{-ip_x x}$, it can be easily derived that

$$Dp^6 - Qp^4 + (1 - \frac{\rho_i d}{\rho g}\omega^2)p^2 = \frac{\omega^2}{gh}, \qquad (6.22)$$

where $p_x^2 + p_z^2 = p_0^2$ with $p_x = p_0 \cos\theta$ and $p_z = p_0 \sin\theta$ and p_0 being the real and positive root of Eq. (6.22). Eq. (6.22) is known as the dispersion relation associated with flexural gravity waves in the presence of uniform compressive force under shallow water approximation, which is also referred to as the linearised flexural gravity long wave dispersion relation. As discussed in Chapter 1, assuming $\rho_i d\omega^2/\rho << 1$ is small enough compared with the other terms and thus neglecting this term, Eq. (6.22) yields

$$(Dp^4 - Qp^2 + 1)p^2 = \omega^2/gh. \qquad (6.23)$$

Thus, Eq. (6.23) is the modified dispersion relation associated with the flexural gravity wave under shallow water approximation under the assumption $\rho_i d\omega^2/\rho << 1$. Therefore, the phase and group velocities associated with the linearised long flexural gravity wave in the plate-covered region are given by

$$c = \sqrt{(Dp^4 - Qp^2 + 1)gh}, \quad c_g = nc, \qquad (6.24)$$

where

$$n = \frac{1}{2}\left\{\frac{5Dp^4 - 3Qp^2 + 1}{Dp^4 - Qp^2 + 1} + 1\right\}.$$

NB. In the case of variable external pressure $P(x, z, t)$ acting on the surface of the plate, Eq. (6.17) is modified as

$$\left\{EI\nabla_{xz}^4 + N\nabla_{xz}^2 + \rho_i d\frac{\partial^2}{\partial t^2} + \rho g\right\}\zeta = \rho\frac{\partial\Phi}{\partial t} - P(x, z, t), \qquad (6.25)$$

and accordingly the long wave equation will be modified. On the other hand, if the beam is of heterogeneous characteristics, then Eq. (6.17) is replaced by

$$\left\{\nabla_{xz}^2\left(\tilde{D}(x, z, t)\nabla_{xz}^2\right) + N\nabla_{xz}^2 + m(x)\frac{\partial^2}{\partial t^2} + \rho g\right\}\zeta = \rho\frac{\partial\Phi}{\partial t}, \qquad (6.26)$$

where $\tilde{D}(x, z, t)$ is called the cylindrical stiffness of the beam and m_s is the specific mass of the beam, and it is assumed that $D(x, z, t)$ and its first derivatives are piecewise continuous and the second derivatives are integrable (as in [133]). Further, it may be noted that the linearised long wave equation in the open water region and plate-covered region is independent of y and depends on the water depth only.

Next, the wave structure interaction problem in the case of a two-layer fluid having an elastic plate-covered surface and an interface is discussed under shallow water assumption. In polar regions, the two-layer fluid is formed due to the melting of the ice sheet as discussed in [123]. For simplicity, it is assumed that the two fluids are immiscible and of density ρ_1 and ρ_2 ($\rho_1 < \rho_2$) and the interface is at the depth h from the mean free surface which is covered by a thin elastic plate as discussed in the case of single-layer fluid. Thus, the upper and lower fluids occupy the regions $-\infty < x, z < \infty$, $0 < y < h$ and $-\infty < x, z < \infty$, $h < y < H$, respectively. It is assumed that the two fluids are inviscid and incompressible and that motion is irrotational with the same angular frequency ω. Thus, there exist velocity potentials $\Phi_j(x, y, z, t)$, $j = 1, 2$, where 1 and 2 are used to indicate the velocity potentials for upper and lower layer fluids, respectively. The velocity potentials $\Phi_j(x, y, z, t)$ satisfy the three-dimensional Laplace equation in terms of the space variables. It is assumed that the elevation of the ice-covered surface and interface are $\zeta_1(x, z, t)$ and $\zeta_2(x, z, t)$, respectively.

Proceeding in a similar manner as in the case of a single-layer fluid, integrating the continuity equation over the interval $(h + \zeta_2, \zeta_1)$ and using the kinematic condition, the continuity equation associated with the linearised long wave in the upper layer fluid yields

$$h\left(\frac{\partial^2 \Phi_1}{\partial x^2} + \frac{\partial^2 \Phi_1}{\partial z^2}\right) = \frac{\partial \zeta_2}{\partial t} - \frac{\partial \zeta_1}{\partial t}. \tag{6.27}$$

Further, integrating the continuity equation over the interval $(h + \zeta_2, H)$ and using the kinematic condition, the continuity equation associated with the linearised long wave in the lower layer fluid yields

$$(H - h)\left(\frac{\partial^2 \Phi_2}{\partial x^2} + \frac{\partial^2 \Phi_2}{\partial z^2}\right) = \frac{\partial \zeta_2}{\partial t}. \tag{6.28}$$

From the elastic plate in Eq. (6.16), in the presence of compressive force, the elastic plate equation yields

$$\left\{EI\nabla_{xz}^4 + N\nabla_{xz}^2 + \rho_i d\frac{\partial^2}{\partial t^2}\right\}\zeta_1 = P_{atm} - P_H, \tag{6.29}$$

where P_{atm} is the constant atmospheric pressure assumed to be zero without loss of generality and P_H is the hydrodynamic pressure acting on the plate and is given by

$$P_H = -\rho_1\frac{\partial \Phi_1}{\partial t} + \rho_1 g\zeta_1. \tag{6.30}$$

Thus, Eqs. (6.29) and (6.30) yield

$$\left(EI\nabla_{xz}^4 + N\nabla_{xz}^2 + \rho_i d\frac{\partial^2}{\partial t^2} + \rho g\right)\zeta_1 = \rho_1\frac{\partial \Phi_1}{\partial t}. \tag{6.31}$$

Using Eq. (6.27), Eq. (6.31) can be rewritten as

$$\left\{D\nabla^6_{xz} + Q\nabla^4_{xz} + \left(\frac{\rho_i d}{\rho g}\frac{\partial^2}{\partial t^2} + 1\right)\nabla^2_{xz}\right\}\zeta_1 = \frac{\partial^2\zeta_2}{\partial t^2} - \frac{\partial^2\zeta_1}{\partial t^2}, \tag{6.32}$$

where D and Q are the same as in Eq. (6.17). Further, at the interface boundary, the continuity of pressure yields

$$s\left(\frac{\partial\Phi_1}{\partial t} - g\zeta_2\right) = \frac{\partial\Phi_2}{\partial t} - g\zeta_2, \quad \text{on} \quad y = h, \tag{6.33}$$

where $s = \rho_1/\rho_2$. Eliminating Φ_1 and Φ_2 from Eqs. (6.27), (6.28) and (6.29), it can be easily derived that

$$\frac{h}{H-h}\frac{\partial^2\zeta_2}{\partial t^2} - s\left(\frac{\partial^2\zeta_2}{\partial t^2} - \frac{\partial^2\zeta_1}{\partial t^2}\right) = (1-s)gh\nabla^2_{xz}\zeta_2. \tag{6.34}$$

Thus, Eqs. (6.32) and (6.34) yield the equation of linearised long wave in the plate-covered region in a two-layer fluid of finite water depth.

Assuming $\zeta_1(x, z, t) = H_1 e^{i(p_x x + p_z z - \omega t)}$ and $\zeta_2(x, z, t) = H_2 e^{i(p_x x + p_z z - \omega t)}$, it can be easily derived that

$$(Dp^4 - Qp^2 + 1 - \rho_i d\omega^2/\rho g)ghp^2 = \omega^2(1 - H_2/H_1) \tag{6.35}$$

and

$$\omega^2 = \frac{(1-s)ghp^2(H-h)H_2}{hH_2 - s(H_1 - H_2)(H-h)}, \tag{6.36}$$

where $p_x^2 + p_z^2 = p^2$. From Eqs. (6.35) and (6.36), eliminating H_2/H_1, the flexural gravity wave dispersion relation in the two-layer fluid under shallow water approximation is obtained as

$$R\omega^4 + S\omega^2 + T = 0, \tag{6.37}$$

where $R = 1$, $S = (Dp^4 - Qp^2 + 1)\{s(H-h) + h\} + (1-s)(H-h)ghp^2$ and $T = (Dp^4 - Qp^2 + 1)(H-h)(1-s)gh^2p^4$.

6.2 Surface gravity wave interaction with floating elastic plates

In the present section, gravity wave interaction with floating flexible plates is analysed under shallow water approximation in several cases assuming that the motion is simple harmonic in time.

6.2.1   Wave scattering by a semi-infinite floating elastic plate

In this subsection, oblique surface wave scattering by a semi-infinite floating elastic plate is studied. The geometry of the physical problem and the characteristic of the fluid motion and elastic plate are assumed to be the same as discussed in Section 6.1. The problem is considered in three dimensions with the $x-z$ plane being the horizontal plane and the y-axis being vertically downward positive, extending from the mean free surface $y = 0$ to the uniform seafloor $y = h$. The fluid occupies the domain $-\infty < x < \infty$, $-\infty < z < \infty$, $0 < y < h$. The region $-\infty < x < 0$, $-\infty < z < \infty$, $0 < y < h$ (referred to as region 1) is assumed to be the open water region, and the region $0 < x < \infty$, $-\infty < z < \infty$, $0 < y < h$ (referred to as region 2) is the plate-covered region with a semi-infinite floating elastic plate of thickness d occupying the surface $0 < x < \infty$, $-\infty < z < \infty$, $y = 0$ as in Figure 6.1. It is assumed that an obliquely incident gravity wave of the form $\zeta_0(x, z, t) = Re(e^{i(k_x x + k_z z - \omega t)})$ impinges on a semi-infinite floating elastic plate at $x = 0$. To maintain geometrical symmetry, the velocity potential $\Phi_j(x, y, z, t)$, the water surface elevation $\zeta_1(x, z, t)$ and deflection of the elastic plate $\zeta_2(x, z, t)$ are assumed to be of the form $\Phi_j(x, y, z, t) = Re\{\phi_j(x, y)e^{ik_z z - i\omega t}\}$, $\zeta_j(x, z, t) = Re\{\eta_j(x)e^{ik_z z - i\omega t}\}$, where Re denotes the real part, ω is the angular frequency and k_z is the z component of the wave number. In the open water region, the equation of continuity for the linearised long wave (as in Section 6.1) is given by

$$-i\omega\eta_1 = h\mathcal{L}(\partial_x)\phi_1, \qquad (6.38)$$

where $\mathcal{L}(\partial_x) = \partial_x^2 - k_z^2$. Further, from Eq. (6.10), the linearised long wave equation of motion in the open water region in terms of the spatial velocity potential is obtained as

$$\left\{\mathcal{L}(\partial_x) + \frac{\omega^2}{gh}\right\}\phi_1 = 0. \qquad (6.39)$$

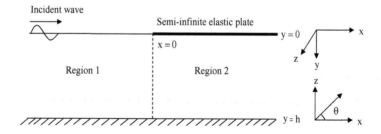

FIGURE 6.1
Schematic diagram of wave scattering by a semi-infinite floating elastic plate.

Further, the linearised long wave equation of motion in the plate-covered regions is given by (as in Eq. (6.19))

$$\left\{ EI\mathcal{L}^3(\partial_x) + N\mathcal{L}^2(\partial_x) + (\rho g - m_s \omega^2)\mathcal{L}(\partial_x) + \frac{\rho \omega^2}{h} \right\} \phi_2 = 0, \qquad (6.40)$$

where $m_s = \rho_i d$. The continuity of energy and the mass flux at the interface $x = 0$ yields

$$\phi_1 = \phi_2 \quad \text{and} \quad \phi_{1x} = \phi_{2x}. \qquad (6.41)$$

Assuming that the semi-infinite plate has a free edge at $x = 0$, the vanishing of bending moment and shear force yield

$$EI(\partial_x^2 - \nu k_z^2)\partial_x^2 \phi_2(0,0) = 0 \qquad (6.42)$$

and

$$\left[EI\{\partial_x^2 - (2 - \nu)k_z^2\} + N \right]\partial_x^3 \phi_2(0,0) = 0. \qquad (6.43)$$

The far field conditions associated with the wave scattering by the semi-infinite plate are given by

$$\phi_j(x,y) = \begin{cases} e^{ik_x x} + R_0 e^{-ik_x x} & \text{as} \quad x \to -\infty, \quad \text{for } j = 1, \\ T_0 e^{ip_{0x} x} & \text{as} \quad x \to \infty, \quad \text{for } j = 2, \end{cases} \qquad (6.44)$$

where R_0 and T_0 are the complex amplitudes of the reflected and transmitted waves, k_0 is the wave number of the plane gravity wave and p_0 is the wave number of the plane flexural gravity wave associated with the linearised long wave propagating below the plate-covered region. Further, $p_{0x} = \sqrt{p_0^2 - k_z^2}$ is the component of the plane progressive flexural gravity wave propagating along the x-axis and k_x is the component of the plane progressive wave propagating along the x-axis in the open water region. Further, it is assumed that $p_{0x} > 0$ for the existence of a plane progressive wave. Thus, the spatial velocity potentials ϕ_j, $j = 1, 2$ associated with the scattering of the plane progressive wave by the semi-infinite floating elastic plate in the respective domains are given by

$$\phi_1(x,y) = e^{ik_x x} + R_0 e^{-ik_x x}, \qquad \text{in region 1,}$$

$$\phi_2(x,y) = T_0 e^{ip_{0x} x} + \sum_{n=I}^{II} T_n e^{i\epsilon_n p_{nx} x}, \quad \text{in region 2,} \qquad (6.45)$$

where $\epsilon_I = 1$ and $\epsilon_{II} = -1$, R_0 and T_0 are associated with the unknown amplitude of the reflected and transmitted waves and $p_{nx} = \sqrt{p_n^2 - k_z^2}, n = 0, I, ..., IV$. The wave number k_0 is the positive real root of the linearised long wave dispersion relation given by

$$k_0^2 = \omega^2 / gh, \qquad (6.46)$$

whereas p_n, $n = 0, I, II$ are the roots of the long wave dispersion relation associated with the flexural gravity waves and are given by

$$Dp_n^6 - Qp_n^4 + (1 - m_s\omega^2/\rho g)p_n^2 = \omega^2/gh. \qquad (6.47)$$

It may be noted that Eq. (6.46) does not have any other root except the two real roots $\pm k_0$, which represent the propagating modes. On the other hand, the plate-covered dispersion relation for shallow water as in Eq. (6.47) has four complex roots p_n, $n = I, II, III, IV$ of the form $\pm a \pm ib$ apart from the two real roots $\pm p_0$. The real roots $\pm p_0$ represent the propagating modes, while the complex roots p_n, $n = I, II, III, IV$ represent the nonpropagating modes that decay at the far field. In the present study, only the positive real roots k_0 and p_0 and the two complex roots p_n, $n = I, II$ having positive real parts with $p_{II} = \bar{p}_I$ are taken into account for the sake of boundedness of the solution, where the bar refers to the complex conjugate. Utilising the continuity of energy and mass flux as in Eq. (6.41) for $j = 1$ along with the free edge conditions in Eqs. (6.42) and (6.43), a system of four linear algebraic equations for the determination of the four unknowns R_0 and T_n, $n = 0, I, II$ are obtained. Once the unknown constants R_0 and T_0 are obtained, the reflection and transmission coefficients K_r and K_t are evaluated directly from the definition as given by

$$K_r = |R_0|, \quad K_t = \left| p_{0x}^2 T_0/k_x^2 \right|. \qquad (6.48)$$

Utilising the law of conservation of energy flux, it can be derived that K_r and K_t satisfy the energy relation

$$K_r^2 + \chi K_t^2 = 1, \qquad (6.49)$$

where $\chi = k_0(\rho g - m_s\omega^2 + 2Np_0^2 + 3EIp_0^4)/p_0\rho g$. It may be noted that the energy relation in Eq. (6.49) can be derived alternately by using Green's theorem. In the next subsection, the solution procedure associated with gravity wave scattering by a finite floating elastic plate under shallow water approximation will be discussed.

6.2.2   Surface wave scattering by a finite floating elastic plate

The problem of the scattering of surface waves by a finite floating elastic plate in uniform water depth h is studied under shallow water approximation. It is assumed that the plate of finite length a and infinite width occupies the region $0 < x < a, -\infty < z < \infty$, $y = 0$. Thus, the whole domain is divided into three subdomains. Region 1 corresponds to the open water region on the side in which the wave is incident on the floating plate, region 2 corresponds to the plate-covered region and region 3 corresponds to the open water region in which the transmitted wave propagates as in Figure 6.2. The assumptions on the fluid motion along with the plate characteristics are the same as described

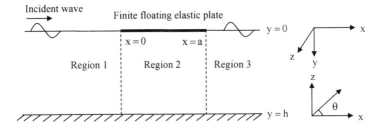

FIGURE 6.2
Schematic diagram of wave scattering by a finite floating elastic plate.

in Subsection 6.2.1. It is assumed that an obliquely incident gravity wave of the form $\zeta_0(x, z, t) = Re(e^{i(k_x x + k_z z - \omega t)})$ impinges on the finite floating elastic plate of length a at $x = a$, with k_x and k_z being the same as defined in Section 6.1. To maintain geometrical symmetry, the velocity potential $\Phi_j(x, y, z, t)$ and the water surface elevation $\zeta_1(x, z, t)$ and deflection of the elastic plate $\zeta_2(x, z, t)$ are assumed to be of the form $\Phi_j(x, y, z, t) = Re\{\phi_j(x, y)e^{ik_z z - i\omega t}\}$, $\zeta_j(x, z, t) = Re\{\eta_j(x)e^{ik_z z - i\omega t}\}$, where Re denotes the real part, ω is the angular frequency and k_z is the z component of the wave number. Thus, in the open water region, the equation of continuity for linearised long wave (as in Section 6.1) yields

$$-i\omega \eta_j = h\mathcal{L}(\partial_x)\phi_j, \quad j = 1, 3, \tag{6.50}$$

where $\mathcal{L}(\partial_x) = \partial_x^2 - k_z^2$. Further, from Eq. (6.12), the linearised long wave equation of motion in the open water region in terms of the spatial velocity potential is obtained as

$$\left\{\mathcal{L}(\partial_x) + \frac{\omega^2}{gh}\right\}\phi_j = 0, \quad j = 1, 3. \tag{6.51}$$

Further, from Eq. (6.19), the linearised long wave equation of motion in the plate-covered region is obtained as

$$\left\{EI\mathcal{L}^3(\partial_x) + Q\mathcal{L}^2(\partial_x) + (\rho g - m_s \omega^2)\mathcal{L}(\partial_x) + \frac{\rho \omega^2}{h}\right\}\phi_2 = 0. \tag{6.52}$$

The continuity of energy and mass flux at the interface $x = 0, a$, yields

$$\phi_1 = \phi_2 \quad \text{and} \quad \phi_{1x} = \phi_{2x}, \quad \text{at} \quad x = 0 \tag{6.53}$$

and

$$\phi_2 = \phi_3 \quad \text{and} \quad \phi_{2x} = \phi_{3x}, \quad \text{at} \quad x = a. \tag{6.54}$$

Assuming that the finite plate has free edges at $(0, 0)$ and $(a, 0)$, vanishing of bending moment and shear force yield

$$EI(\partial_x^2 - \nu k_z^2)\partial_x^2 \phi_2(c, 0) = 0 \tag{6.55}$$

and
$$[EI\{\partial_x^2 - (2-\nu)k_z^2\} + N]\partial_x^3 \phi_2(c,0) = 0, \qquad (6.56)$$

where $c = 0, a$. The far field conditions associated with the wave scattering by the semi-infinite plate are given by

$$\phi_j(x,y) = \begin{cases} e^{ik_x x} + R_1 e^{-ik_x x} & \text{as} \quad x \to -\infty, \quad \text{for} \quad j=1, \\ T_1 e^{ik_x x} & \text{as} \quad x \to \infty, \quad \text{for} \quad j=3, \end{cases} \qquad (6.57)$$

where R_1 and T_1 are the complex amplitudes of the reflected and transmitted waves, k_0 is the wave number of the plane progressive gravity wave associated with the linearised long wave and k_x is the same as defined in Section 6.1.

The velocity potentials ϕ_j for $j = 1, 2, 3$ in the respective domains are expressed as

$$\phi_1(x,y) = e^{ik_x x} + R_1 e^{-ik_x x}, \text{ for } x < 0,$$
$$\phi_2(x,y) = A_1 e^{ip_{0x} x} + B_1 e^{-ip_{0x} x} + \sum_{n=I}^{IV} A_n e^{-ip_{nx} x}, \text{ for } 0 < x < a, \qquad (6.58)$$
$$\phi_3(x,y) = T_1 e^{ik_x x}, \text{ for } x > a,$$

where $p_{nx} = \sqrt{p_n^2 - k_z^2}, n = 0, I, ..., IV$. Utilising continuity of energy and mass flux at $x = 0, a$ as in Eqs. (6.53) and (6.54) for $j = 1, 2$, the free edge conditions as in Eqs. (6.55) and (6.56) determine the eight unknowns $R_1, A_1, B_1, A_n, n = I, II, III, IV$ and T_1. Once the unknown constants R_1 and T_1 are obtained, the reflection and transmission coefficients K_r and K_t are evaluated directly, as given by

$$K_r = |R_1|, \quad K_t = |T_1|. \qquad (6.59)$$

Further, it may be noted that the reflection and transmission coefficients satisfy the energy relation
$$K_r^2 + K_t^2 = 1, \qquad (6.60)$$
which is often used to check the accuracy of the computation.

Apart from the direct method of solution, from the known results of reflection and transmission coefficients associated with the wave scattering by a semi-infinite floating elastic plate, the reflection and transmission coefficients associated with the wave scattering by a finite floating elastic plate can be obtained using the wide spacing approximation method as discussed in [32] [60] and [146]. In the case of wave scattering by a finite floating elastic plate, the gravity wave experiences partial reflection and transmission by the floating elastic plate at $x = 0$ and $x = a$. It is assumed that the plate length is much larger than the wave length of the incident plane progressive wave, that is, $|a| >> \lambda$ with λ being the incident wavelength (as in [29], [92]) to ensure that the evanescent modes do not contribute to the solution. Thus, the local effects produced during the interaction of the incident waves with one of the plate edges do not affect the interaction at the other edges of the plate. Thus, the asymptotic forms of the velocity potential ϕ_j for $j = 1, 2, 3$ are given by

$$
\begin{aligned}
\phi_1 &\sim e^{ik_x x} + R_1 e^{-ik_x x}, \quad -\infty < x < 0, \\
\phi_2 &\sim A_1 e^{ip_{0x} x} + B_1 e^{-ip_{0x} x}, \ 0 < x < a, \\
\phi_3 &\sim T_1 e^{ik_x x}, \quad a < x < \infty.
\end{aligned}
\tag{6.61}
$$

Before applying the wide spacing approximation method, it may be noted that R_0 and T_0 correspond to the amplitude of the reflected and transmitted waves associated with the wave scattering by a semi-infinite floating elastic plate at $x = 0$ in isolation satisfying the radiation condition

$$
\phi(x, y) = \begin{cases}
e^{ik_x x} + R_0 e^{-ik_x x} & \text{as} \quad x \to -\infty, \\
T_0 e^{ik_x x} & \text{as} \quad x \to \infty,
\end{cases}
\tag{6.62}
$$

as discussed in the previous subsection. Further, it is assumed that r_1 and t_1 correspond to the amplitude of the reflected and transmitted waves associated with the wave scattering by a semi-infinite plate at $x = a$ in isolation satisfying the radiation condition

$$
\phi(x, y) = \begin{cases}
e^{-ik_x x} + r_1 e^{ik_x x} & \text{as} \quad x \to \infty, \\
t_1 e^{-ik_x x} & \text{as} \quad x \to -\infty.
\end{cases}
\tag{6.63}
$$

Now, to determine the unknown coefficients in Eq. (6.61) associated with wave scattering by a finite floating elastic plate, equating the left and right components of the propagating waves at the plate edges $x = 0, a$ with the amplitudes of the reflected and transmitted waves in the prescribed region (as in Figure 6.3), a system of 4 linear equations associated with 4 unknowns R_1, T_1, A_1, B_1 is obtained as (as in [32])

$$
\begin{aligned}
R_1 &= R_0 + B_1 t_1, \\
A_1 e^{ip_{0x} a} &= B_1 r_1 e^{-ip_{0x} a} + T_0 e^{ik_x a}, \\
B_1 e^{-ip_{0x} a} &= A_1 r_1 e^{ip_{0x} a}, \\
T_1 e^{ik_x a} &= A_1 t_1 e^{ip_{0x} a}.
\end{aligned}
\tag{6.64}
$$

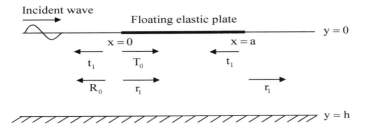

FIGURE 6.3
Schematic diagram of wave direction in wide spacing approximation method.

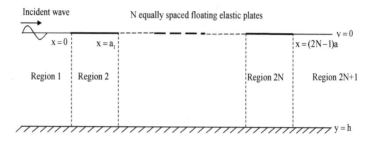

FIGURE 6.4
Schematic diagram of wave scattering by multiple floating elastic plates.

Thus, solving the above system of equations, the reflection and transmission coefficients $K_r = |R_1|$ and $K_t = |T_1|$ associated with the gravity wave scattering by the finite floating elastic plate is derived.

6.2.3   Wave scattering by multiple floating elastic plates

In this subsection, wave scattering by a finite number of floating elastic plates is analysed as in Figure 6.4. It is assumed that there are N plates which occupy the regions $\cup I_j$ where $I_j = ((2j-2)a, (2j-1)a)$ for $j = 1, 2, ..., N$ along the x-axis with $-\infty < z < \infty$ and $0 < y < h$ and they are referred as the plate-covered regions. Further, the gaps occupy the open water regions $\cup G_j$ with $G_j = ((2j-1)a, 2ja)$ for $1 = 2, 3, ..., N-1$, $(-\infty, 0)$ and $((2N-1)a, \infty)$ on the free surface along the x-axis with $-\infty < z < \infty$ and $0 < y < h$. The physical assumption remains the same as discussed in the previous subsection. Further, in the open water region, the spatial velocity potentials ϕ_j satisfy the long wave equation in Eq. (6.51) and in the plate-covered region the velocity potential ϕ_j satisfies the long wave equation as in Eq. (6.52). The continuity of energy and mass flux at the interface $x = ja$, $j = 0, 1, 2, ..., 2N-1$ yields

$$\phi_{j+1} = \phi_j \quad \text{and} \quad \frac{\partial \phi_{j+1}}{\partial x} = \frac{\partial \phi_j}{\partial x}. \tag{6.65}$$

Assuming that the floating elastic plates have free edges, the bending moment and shear force are assumed to be zero at the end points $x = ja$, $j = 1, ..., 2N$, which yields

$$EI(\partial_x^2 - \nu k_z^2)\partial_x^2 \phi_j(x, 0) = 0 \tag{6.66}$$

and

$$\left[EI\{\partial_x^2 - (2-\nu)k_z^2\} + N\right]\partial_x^3 \phi_j(x, 0) = 0. \tag{6.67}$$

The far field conditions are given by

$$\phi_j(x, y) = \begin{cases} e^{ik_x x} + R_N e^{-ik_x x} & \text{as} \quad x \to -\infty \quad \text{for } j = 1 \\ T_N e^{ik_x x} & \text{as} \quad x \to \infty \text{ for } j = 2N+1, \end{cases} \tag{6.68}$$

where R_N and T_N are the complex amplitudes of the reflected and transmitted waves associated with the scattering of waves in the presence of N plates. The wave numbers k_x and k_z are the same as defined in Section 6.1. The velocity potentials $\phi_j(x, y)$, for $j = 1, 2, ..., 2N + 1$ are expanded as

$$
\phi_j(x, y) = \begin{cases}
e^{ik_x x} + R_N e^{-ik_x x}, & \text{for} \quad -\infty < x < 0, \\
\bar{A}_1^{2j-1} e^{ip_0 x} + \overset{IV}{\underset{n=0,I}{\sum}} A_n^{2j-1} e^{-ip_{nx} x}, & \text{for} \quad x \in I_j, \\
A_1^{2j-2} e^{ik_x x} + B_1^{2j-2} e^{-ik_x x}, & \text{for} \quad x \in G_j, \\
T_N e^{ik_x x}, & \text{for} \quad x > 2Na.
\end{cases} \tag{6.69}
$$

It may be noted that the k_x is the same as defined in Section 6.1, and p_{nx} is the same as in Eq. (6.58). Using the continuity of energy and mass flux as in Eq. (6.65) and the free edge condition at $x = ja$ for $j = 0, 1, ..., 2N - 1$, a linear system of equations is obtained for the determination of the unknowns. Once the unknowns R_N and T_N are obtained, the reflection and transmission coefficients are derived from the relations $K_r = |R_N|$ and $K_t = |T_N|$.

Next, the wide spacing approximation method is discussed to analyse the gravity wave interaction with multiple plates. The procedure is similar to the one discussed in [32]. In the case of wave interaction with multiple floating elastic plates, the gravity wave experience partial reflection and the transmission at each plate edges $x = ja$ for $j = 0, 1, 2, ..., 2N-1$, it is assumed that the distance between two consecutive plates is much larger than the wave length of the incident plane progressive wave, that is, $|a| \gg \lambda$ with λ being the incident wave length (as in [29], [92]) to ensure that the evanescent modes do not contribute to the solution. Thus, the local effects produced during the interaction of the incident wave with the individual plates do not affect the subsequent interactions. Assuming that the plates are placed wide apart and each of the plates is large in length, the asymptotic forms of the velocity potential ϕ_j for $j = 1, 2, ..., 2N+1$ far away from the plate edges in the respective regions are given by

$$
\begin{aligned}
\phi_1 &\sim e^{ik_x x} + R_N e^{-ik_x x}, & -\infty < x < 0, \\
\phi_{2j-1} &\sim A_{2j-1} e^{ip_x x} + B_{2j-1} e^{-ip_x x}, & x \in I_j, \ j = 1, 2, ..., N-1, \\
\phi_{2j-2} &\sim A_{2j-2} e^{ik_x x} + B_{2j-2} e^{-ik_x x}, & x \in G_j, \ j = 2, ..., N+1, \\
\phi_{N+1} &\sim T_N e^{ik_x x}, & x > (2N - 1)a.
\end{aligned} \tag{6.70}
$$

Equating the left and right components of the propagating waves at the plate edges $x = ja$ for $j = 0, 1, 2, ..., 2N - 1$ with the amplitudes of the reflected and transmitted waves in the prescribed region, a system of linear equations associated with the unknowns R_N, T_N, A_j, B_j as in Eq. (6.70) is obtained as given by

$$
\begin{aligned}
R_N &= R_0 + B_1 t_1, \\
B_{2j-1} e^{-ik_x(2j-1)a} &= r_{2j-1} A_{2j-1} e^{ik_x(2j-1)a} + B_{2j-2} t_{2j-2} e^{-ip_x(2j-1)a}, \\
A_{2j-2} e^{ip_x(2j-2)a} &= r_{2j-2} B_{2j-2} e^{-ip_x(2j-2)a} + t_{2j-1} A_{2j-1} e^{ik_x(2j-2)a}, \\
T_N e^{ik_x(2N-1)a} &= A_{2N-1} t_{2N-1} e^{ik_x(2N-1)a},
\end{aligned}
$$

$$(6.71)$$

Here, r_j and t_j correspond to the amplitude of the reflected and transmitted waves by a single floating elastic plate at $x = ja$, $j = 1, 2, 3, ..., (2N-1)$ in isolation. Further, R_0 and T_0 are the reflection and transmission coefficients associated with a single plate at $x = 0$ in isolation. Thus, solving the above system of equations, the reflection and transmission coefficients $K_r = |R_N|$ and $K_t = |T_N|$ for N floating plates are obtained.

6.3 Flexural gravity wave scattering by articulated floating plates

In the previous section, surface gravity wave interaction with floating elastic plates is discussed under shallow water approximation. In the present section, emphasis is on flexural gravity wave scattering due to articulated plates. Both the cases of single and multiple articulation are discussed.

6.3.1 Effect of articulation on two semi-infinite floating plates

In the present subsection, wave scattering by two semi-infinitely extended articulated floating elastic plates is studied. In the polar region, partially frozen cracks in floating ice sheets are modeled as articulated floating elastic plates (see [104]). It is assumed that an obliquely incident flexural gravity wave of the form $\zeta_0(x, z, t) = Re(e^{i(p_x x + p_z z - \omega t)})$ impinges the semi-infinite floating elastic plate at $x = 0$ as in Figure 6.5, with p_x and p_z being the same as defined in Section 6.1. To maintain geometrical symmetry, the velocity potentials $\Phi_j(x, y, z, t)$ and the water surface elevation $\zeta_1(x, z, t)$ and deflection of the elastic plate $\zeta_2(x, z, t)$ are assumed to be of the form $\Phi_j(x, y, z, t) = Re\{\phi_j(x, y)e^{ip_z z - i\omega t}\}$, $\zeta_j(x, z, t) = Re\{\eta_j(x)e^{ip_z z - i\omega t}\}$, where Re denotes the real part, ω is the angular frequency, p_z is the z component of the wave number and $j = 1, 2$. Under the assumption of small amplitude structural response and linearised shallow water theory as discussed in the previous subsections, flexural gravity wave scattering by the articulated floating structure is studied by determining the velocity potentials ϕ_j which satisfy the linearised long wave equation as in Eq. (6.21) in the regions $-\infty < x < 0$ and $0 < x < \infty$ in the horizontal direction along the x-axis with $-\infty < z < \infty$

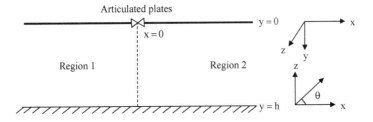

FIGURE 6.5
Schematic diagram of wave scattering due to single articulation.

and $0 < y < h$. Further, at the interface $x = 0$, ϕ_j satisfies the matching conditions in Eq. (6.41). Further, the two semi-infinite floating ice sheets satisfying the articulated edge conditions at $x = 0$ are given by

$$S\left(\frac{\partial}{\partial x}\right)\phi_{j+1}(x+,0) = -k_{55}\partial_x^2\left\{\phi_{j+1}(x+,0) - \phi_j(x-,0)\right\}$$

$$S\left(\frac{\partial}{\partial x}\right)\phi_j(x-,0) = -k_{55}\partial_x^2\left\{\phi_{j+1}(x+,0) - \phi_j(x-,0)\right\} \qquad (6.72)$$

and

$$EI(\partial_x^2 - \nu p_z^2)\partial_x^2\phi_{j+1}(x+,0) = k_{33}\partial_x^3\left\{\phi_{j+1}(x+,0) - \phi_j(x-,0)\right\},$$

$$EI(\partial_x^2 - \nu p_z^2)\partial_x^2\phi_j(x-,0) = k_{33}\partial_x^3\left\{\phi_{j+1}(x+,0) - \phi_j(x-,0)\right\}, \qquad (6.73)$$

where $S\left(\frac{\partial}{\partial x}\right) = \left[EI\{\partial_x^2 - (2-\nu)p_z^2\} + N\right]\partial_x^3$. The far field conditions associated with the wave scattering by the semi-infinite articulated plates are given by

$$\phi_j(x,y) = \begin{cases} e^{i\mu_0 x} + R_0 e^{-i\mu_0 x} & \text{as } x \to -\infty, \quad \text{for } j = 1, \\ T_0 e^{i\mu_0 x} & \text{as } x \to \infty, \quad \text{for } j = 2, \end{cases} \qquad (6.74)$$

with R_0 and T_0 being the complex amplitudes of the reflected and transmitted waves and p_0 being the wave number of the plane progressive flexural gravity wave associated with the linearised long flexural gravity wave and $\mu_0 = \sqrt{p_0^2 - p_z^2}$. The spatial velocity potentials ϕ_j, $j = 1, 2$ are expressed in the respective domains as given by

$$\phi_1(x,y) = e^{i\mu_0 x} + R_0 e^{-i\mu_x x} + \sum_{n=I}^{II} R_n e^{i\epsilon_n \mu_n x}, \quad \text{in region 1},$$

$$\phi_2(x,y) = T_0 e^{i\mu_0 x} + \sum_{n=I}^{II} T_n e^{\epsilon_n i\mu_n x}, \qquad \text{in region 2}, \qquad (6.75)$$

where $\epsilon_I = 1$ and $\epsilon_{II} = -1$, R_0 and T_0 are associated with the unknown amplitudes of the reflected and transmitted waves and R_n and T_n are the amplitudes of the nonpropagating modes to be determined. Further, in Eq. (6.75), μ_n is given by $\mu_n = \sqrt{p_n^2 - p_z^2}$, $n = 0, I, II$. Using the continuity conditions in Eq. (6.41) and the articulated edge conditions in Eqs. (6.72) and (6.73), from Eqs. (6.75), a system of six equations is derived which is solved to obtain the full solution.

6.3.2   Flexural gravity wave scattering by multiple articulated plates

In the previous subsection, flexural gravity wave scattering due to single articulation is analysed. In the present subsection, flexural gravity wave scattering by multiple articulation is analysed. In this subsection, apart from a brief description on the direct solution approach, the wide spacing approximation method is briefly discussed. In this case, the plates occupy the region I_j along the x-axis in the horizontal plane with $-\infty < x < \infty$ and $0 < y < h$, where I_j is given by $-\infty < x < 0, 0 < x < a, a < x < 2a, ..., (N-1)a < x < \infty$ as in Figure 6.6. The velocity potentials associated with the wave scattering by multiple articulated joints are given by ϕ_j for $j = 1, 2, ..., N+1$. It is obvious that apart from the two semi-infinite plates at the two ends, all the intermediate plates are of finite and equal length. These plates have articulated joints at the edges. Thus, the spatial velocity potential ϕ_j in the plate-covered region satisfies the long wave equation as in Eq. (6.21) and the articulated edge conditions as in Eqs. (6.72) and (6.73) at $x = ja$ for j = 0,1,2,...., $N-1$. Further, at the interface $x = ja$ for $j = 0, 1, 2, ..., N-1$, ϕ_j satisfies the matching conditions as in Eq. (6.41). Finally, the velocity potentials ϕ_j satisfy the far field conditions given by

$$\phi_j(x,y) = \begin{cases} e^{i\mu_0 x} + R_N e^{-i\mu_0 x} & \text{as} \quad x \to -\infty, \quad \text{for } j = 1, \\ T_N e^{i\mu_0 x} & \text{as} \quad x \to \infty, \quad \text{for } j = N+1. \end{cases} \quad (6.76)$$

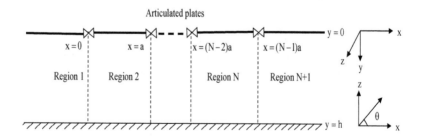

Articulated plates

FIGURE 6.6
Schematic diagram of wave scattering due to multiple articulations.

Thus, the velocity potentials $\phi_j(x,y)$ for $j = 1, 2, ..., N+1$ are expanded as

$$
\phi_j(x,y) = \begin{cases} e^{i\mu_0 x} + R_N e^{-i\mu_0 x} + \sum\limits_{n=I}^{II} R_{Nn} e^{i\epsilon_n \mu_n x}, & \text{for } x < 0, \\[2ex] A_j e^{i\mu_0 x} + B_j e^{-i\mu_0 x} + \sum\limits_{n=I}^{IV} A_n^j e^{-i\mu_n x}, & \text{for } x \in I_j, \\[2ex] T_N e^{i\mu_0 x} + \sum\limits_{n=I}^{II} T_{Nn} e^{i\epsilon_n \mu_n x}, & \text{for } x > (N-1)a, \end{cases}
\tag{6.77}
$$

where $I_j = (ja, (j+1)a)$ with $R_N, T_N, R_{Nn}, T_{Nn}, n = I, II$ and $A_j, B_j, A_n^j, n = I, II, III, IV$ for $j = 2, ..., N$ are the unknown constants to be determined and p_n satisfies the shallow water flexural gravity wave dispersion relation given by

$$
EI p_n^6 - Q p_n^4 + (\rho g - m_s \omega^2) k_n^2 = \rho \omega^2 / h.
\tag{6.78}
$$

Proceeding in a similar manner as in Subsection 6.3.1, using the continuity of energy and mass flux in Eq. (6.41), the edge conditions in Eqs. (6.72)–(6.73) and the far field condition in Eq. (6.76), a system of equations is obtained for determining all the unknowns in the velocity potentials given by Eq. (6.77). Once the unknowns R_{N0} and T_{N0} are obtained, the reflection and transmission coefficients are derived from the relations $K_r = |R_N|$ and $K_t = |T_N|$.

Apart from the direct method of solution discussed above, using the wide spacing approximation method as described in Subsection 6.2.2, wave scattering by multiple articulated floating elastic plates in the case of shallow water approximation is obtained and the details are briefly discussed next. Here, the flexural gravity wave experiences partial reflection and transmission at the articulated joints, located at $x = ja$ for $j = 0, 1, 2, ..., N - 1$. It is assumed that the distance between two consecutive articulated joints is much larger than the wave length of the incident plane progressive wave, that is, $|a| >> \lambda$, with λ being the incident wavelength (as in [29], [92]) to ensure that the evanescent modes do not contribute to the solution. Thus, the local effects produced during the interaction of the incident wave with one of the articulated joints do not affect the subsequent interactions. Assuming that the articulated edges are placed widely apart, the asymptotic forms of the velocity potential ϕ_j for $j = 1, 2, ..., N+1$ far away from the articulations in the respective regions are given by

$$
\begin{aligned}
\phi_1 &\sim e^{i\mu_0 x} + R_N e^{-i\mu_0 x}, \quad x < 0, \\
\phi_{j+1} &\sim A_j e^{i\mu_0 x} + B_j e^{-i\mu_0 x}, \quad (j-1)a < x < ja, \quad j = 1, 2, ..., N-1, \\
\phi_{N+1} &\sim T_N e^{i\mu_0 x}, \quad x > (N-1)a.
\end{aligned}
\tag{6.79}
$$

Here, assuming that R_0 and T_0 are the reflection and transmission coefficients associated with the wave scattering by a single articulation for waves incident upon $x = 0$ as in the previous subsection, the radiation condition is given by

$$\phi(x, y) = \begin{cases} e^{i\mu_0 x} + R_0 e^{-i\mu_0 x} & \text{as } x \to -\infty, \\ T_0 e^{i\mu_0 x} & \text{as } x \to \infty. \end{cases} \tag{6.80}$$

Further, assuming r_{j-1} and t_{j-1} correspond to the reflection and transmission coefficients associated with wave scattering for waves incident upon $x = (j-1)a$, the associated radiation condition is given by

$$\phi(x, y) = \begin{cases} e^{i\mu_0 x} + r_{j-1} e^{-i\mu_0 x} & \text{as } x \to \infty, \\ t_{j-1} e^{i\mu_0 x} & \text{as } x \to -\infty. \end{cases} \tag{6.81}$$

In addition, assuming r_j and t_j are the reflection and transmission coefficients associated with wave scattering for waves incident upon $x = ja$ (wave incident from the right side), the radiation condition is given by

$$\phi(x, y) = \begin{cases} e^{-i\mu_0 x} + r_j e^{i\mu_0 x} & \text{as } x \to \infty, \\ t_j e^{-i\mu_0 x} & \text{as } x \to -\infty. \end{cases} \tag{6.82}$$

Applying the wide spacing approximation as discussed in Section 6.2 (see [32] for further details), a system of equations is obtained as

$$\begin{aligned} R_N &= R_0 + B_1 t_1, \\ B_{j-1} e^{-i\mu_0 (j-1)a} &= r_{j-1} A_{j-1} e^{i\mu_0 (j-1)a} + B_j t_j e^{-i\mu_0 (j-1)a}, \\ A_j e^{i\mu_0 ja} &= r_j B_j e^{-i\mu_0 ja} + t_{j-1} A_{j-1} e^{i\mu_0 ja}, \\ T_N e^{i\mu_0 (N-1)a} &= A_{N-1} t_N e^{i\mu_0 (N-1)a}, \end{aligned} \tag{6.83}$$

with $r_0 = R_0, t_0 = T_0$, $A_0 = 1, B_0 = R_N$ and $j = 1, ..., N - 1$. Here r_j and t_j correspond to the amplitude of the reflected and transmitted waves for single articulation at the articulated edge $x = ja$, $j = 1, 2, ..., N$ in isolation. Further, it may be noted that $r_j = r_{j-1}, t_j = t_{j-1}$ as the medium is homogeneous and the structure is symmetric along the horizontal direction. From Eq. (6.82), it can be easily seen that $A_j, B_j, A_{j-1}, B_{j-1}$ are related by the iterative formula

$$\begin{bmatrix} A_j \\ B_j \end{bmatrix} = \begin{bmatrix} t_{j-1} - (r_{j-1}/t_j)e^{-i\mu_0(j-1)a} & e^{-2i\mu_0 ja}/t_j \\ -(r_{j-1}/t_j)e^{2i\mu_0(j-1)a} & 1/t_j \end{bmatrix} \begin{bmatrix} A_{j-1} \\ B_{j-1} \end{bmatrix}. \tag{6.84}$$

Thus, solving the above system of equations as in Eq. (6.83), the reflection and transmission coefficients $K_r = |R_N|$ and $K_t = |T_N|$ associated with the flexural gravity wave scattering in the presence of N articulated points can be derived. The results derived based on the direct method and the wide spacing approximation method can be compared from the computational results.

6.4 Conclusion

In the present chapter, in addition to deriving the linearised long wave equations to deal with wave interaction with large floating structures, solution approaches for a class of problems are briefly discussed. These results can be generalised to include structural heterogeneity, abrupt changes in bottom topography and fluid stratification. In the wide spacing approximation method discussed in the present chapter, only the contributions from far field are taken in the solution. However, the wide spacing approximation used in the present chapter can be easily generalised to include the evanescent modes by the recently generalised method in [146]. Further details on wave interaction with very large floating structures based on shallow water approximation can be found in [60], [61] and [109]. Apart from the time-harmonic problems, there has been significant progress in the last decade on time-dependent wave structure interaction problems under shallow water approximation. Most of these results are based on the application of the spectral method or the mode decomposition method and details are deferred here and can be found in [66], [95], [96], [131], [133], [134] and [135] and the literature cited therein. Reduction of structural responses on floating structures is of paramount interest due to the wide application of these structures in marine environments. The recent developments on various approaches for reduction of structural responses can be found in [39], [53], [54], [55] and [69]. A large class of problems on wave structure interaction can be analysed based on shallow water approximation for structural response attenuation of very large floating structures. Apart from linear approximation, a large class of problems can be analysed based on nonlinear shallow water approximation (as in [151]).

6.5 Examples and exercises

Exercise 6.1 Derive the linearised long wave equation associated with the interaction of surface gravity wave with a floating circular flexible plate in finite water depth.

Exercise 6.2 Derive the linearised long wave equation in two dimensions associated with the interaction of surface gravity wave with a submerged plate in finite water depth.

Exercise 6.3 Derive the two-dimensional linearised long wave equation associated with wave interaction with large floating structures in the presence of current and thus analyse the particle kinematics.

Exercise 6.4 Discuss the problem of interaction of surface waves with a semi-infinite floating elastic plate in a two-layer fluid having an interface under shallow water approximation in finite water depth.

Exercise 6.5 Discuss the problem of interaction of surface gravity waves with a semi-infinite floating elastic plate in the presence of a semi-infinite submerged flexible plate under the assumption of linearised shallow water approximation.

Exercise 6.6 Discuss the problem of attenuation of structural response of a very large floating structure in the presence of a submerged flexible structure under the assumption of linearised shallow water approximation.

Exercise 6.7 Discuss the problem of transient flexural gravity problem under shallow water approximation in the presence of initial impulse and depression.

Exercise 6.8 Discuss the problem of transient flexural gravity problem under shallow water approximation in the presence of a load moving at the plate-covered free surface at uniform speed.

Exercise 6.9 Discuss the scattering of flexural gravity waves due to an abrupt change in the rigidity of the floating structure under the shallow water approximation.

Exercise 6.10 Discuss the scattering of flexural gravity waves due to an abrupt change in bottom topography under the shallow water approximation.

Exercise 6.11 Discuss the scattering of flexural gravity waves in the presence of a submerged dyke of finite width under the shallow water approximation.

7

Boundary integral equation method

7.1 General introduction

In Chapter 3, expansion formulae for wave structure interaction problems are derived based on suitable utilisation of Green's function. In the present chapter, the use of the boundary integral equation method will be demonstrated by considering the surface wave diffraction by a flexible floating membrane. Using the free surface Green's function, the boundary value problems associated with the wave structure interaction problems are converted to integro-differential equations in the cases of both finite and infinite water depths. The integro-differential equations are reduced to a system of equations by two different approaches. From the solution of the system of equations, various physical quantities of interest associated with the physical problem can be obtained.

7.2 Wave diffraction by a floating membrane

In the last two decades, there has been an enormous interest in the study of surface wave interaction with flexible membranes (both vertical and horizontal floating and submerged structures) for various coastal engineering activities (see [20], [63], [79] and the literature cited therein). One of the robust methods to deal with such problems is based on the boundary integral equation method. A numerical model based on the boundary integral equation method is developed in [144] to analyse the interaction of surface waves with a flexible, floating breakwater consisting of a compliant, beam-like structure anchored to the sea bed and kept under tension by a small buoyancy chamber at the tip. The interaction of surface gravity waves with a tensioned, unstretchable, vertical flexible membrane extended to the seabed is analysed based on the boundary integral equation method in [70]. These results are generalised to study the interaction of oblique incident waves with a flexible membrane wave barrier with the membrane tension provided by a buoy using the boundary integral equation method in [17]. The accuracy and convergence of the method are checked using the energy-conservation formula. Further, the boundary integral method for wave interaction with dual membranes tensioned by buoys is studied in [19]. The effectiveness of a flexible porous barrier for protecting cordgrass seedlings from wave action during the critical initial stages of growth following planting is studied based on the boundary integral equation method in [145]. Additionally, oblique wave interaction with a submerged horizontal membrane by using the mode expansion method and the boundary integral equation method is investigated in [19].

Surface gravity wave diffraction by a flexible floating platform is studied using the boundary element method and the suitable application of the modal function associated with the freely floating plate in [48]. Using the integro-differential equation formulation in [48] and an expansion based on the horizontal eigenfunctions associated with the flexural gravity waves propagating below the floating flexible plate, hydroelastic performance of the flexible floating plate is analysed in water of finite and infinite depths in [2]. The aforementioned boundary element method is generalised to study the effect of structural heterogeneity of a finite floating platform in [49]. Hydroelastic analysis of a circular plate on water of finite and infinite depth is studied in [3]. Further, using the boundary integral equation method, the hydroelastic behaviour of a ring-shaped floating plate is analysed in [4]. The boundary integral method is generalised to deal with a thick flexible dock in [50]. The boundary integral equation method has been applied to deal with the interaction of large amplitude ocean waves with a compliant floating raft such as a sea–ice floe in [128].

The surface gravity wave diffractions by a finite floating membrane are demonstrated in two dimensions in the linearised water wave theory and small amplitude membrane deflection in the cases of both finite and infinite water depths. Utilising the free surface Green's function, the boundary value problem associated with the physical problem is converted into a third-order integro-differential equation in terms of the membrane deflection. To solve the integro-differential equation, two different approaches are used. In the first approach, the membrane deflection is expressed in terms of the associated modal functions as in [48]. In the second approach, the membrane deflection is expressed in terms of the horizontal eigenfunctions associated with the dispersion relation of the membrane-covered water surface (as in [2], [51] and [102]). Once, the membrane deflection is obtained, the reflection and transmission coefficients are obtained from the far field behavior of the scattered velocity potential in a straight forward manner. In the present analysis, it is assumed that the membrane has fixed and spring-supported end conditions.

7.3   Mathematical formulation

Under the assumption of the linearised theory of water waves, the problem is considered in the two-dimensional Cartesian coordinate system with the x-axis being in the horizontal direction, and the y-axis in the vertically downward positive direction. The membrane is assumed to be of homogeneous property and water is of a uniform density ρ. The fluid is assumed to be infinitely extended and occupies the region $-\infty < x < \infty$ along with $0 < y < h$ in the case of finite water depth and $0 < y < \infty$ in the case of infinite water depth. The floating membrane is placed at $y = 0$, $0 < x < l$ at the water surface as in Figure 7.1 where l is the length of the membrane. Thus, the fluid domain is a combination of the open water region and the membrane-covered region. The open water region is referred as region R_1, the membrane-covered region is referred as region R_2 and the interface region is referred as ∂R. Further, the open water mean free surface is denoted as $\mathcal{F}$ and the membrane-covered mean surface is denoted as $\mathcal{M}$. Assuming that the fluid is inviscid and incompressible and that the motion is irrotational and simple harmonic in time with angular frequency ω, there exists a velocity potential $\Phi(x, y, t)$ of the form $\Phi(x, y, t) = Re[\phi(x, y)e^{-i\omega t}]$, which satisfies the two-dimensional Laplace equation as given by

$$\nabla^2 \Phi = 0 \quad \text{in the fluid region.} \tag{7.1}$$

The linearised free surface boundary condition at the mean free surface $y = 0$ is given by

$$\frac{\partial^2 \Phi}{\partial t^2} - g\frac{\partial \Phi}{\partial y} = 0 \quad \text{on} \quad y = 0. \tag{7.2}$$

The dynamic pressure exerted on the flexible membrane is given by

$$T\frac{\partial^2 \zeta}{\partial x^2} + m\frac{\partial^2 \zeta}{\partial t^2} = -P_s, \tag{7.3}$$

where T is the tension acting on the membrane, $m = \rho_i d$ is the density of the flexible membrane, g is the acceleration due to gravity, d is the thickness

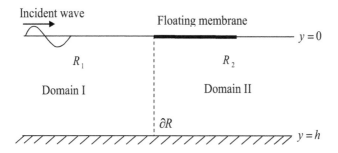

FIGURE 7.1
Schematic diagram of wave diffraction by flexible membrane.

of the flexible membrane and $\zeta(x,t) = Re[\eta(x)e^{-i\omega t}]$ denotes the deflection of the membrane. Further, the hydrodynamic pressure exerted on the flexible membrane at the free surface $y = 0$ is given by

$$P_H = -\rho \frac{\partial \Phi}{\partial t} + \rho g \zeta. \tag{7.4}$$

The linearised kinematic condition on the free surface at $y = 0$ is given by

$$\frac{\partial \zeta}{\partial t} = \frac{\partial \Phi}{\partial y} \quad \text{at} \quad y = 0. \tag{7.5}$$

Assuming $P_s = P_H$ and combining Eqs. (7.3), (7.4) and (7.5), the linearised boundary condition on the membrane-covered surface is obtained as

$$\left\{ \frac{T}{\rho g} \frac{\partial^2}{\partial x^2} + \frac{m}{\rho g} \frac{\partial^2}{\partial t^2} + 1 \right\} \frac{\partial \Phi}{\partial y} - \frac{1}{g} \frac{\partial^2 \Phi}{\partial t^2} = 0 \quad \text{at} \quad y = 0. \tag{7.6}$$

The continuity of velocity and pressure at the interface $x = 0, l$ yields

$$\Phi_x(x+, y) = \Phi_x(x-, y), \ \Phi(x+, y) = \Phi(x-, y), \ \text{at } x = 0, \ l, \ 0 < y < h. \tag{7.7}$$

Further, the edge condition is prescribed at the two end points of the finite membrane. In the present study two types of edges are considered as far as the membrane is concerned, namely, fixed edge and the spring-supported edge. For a flexible membrane having a fixed edge, the transverse displacement vanishes, which yields

$$\zeta(x,t) = 0 \quad \text{at the fixed edges.} \tag{7.8}$$

On the other hand, in case of flexible membranes having a spring-supported edge, it is assumed that a weightless rigid loop is attached at the end of the membrane. The force required to displace the membrane in the case of the spring-supported edge is $k_s \zeta$ where k_s is the spring constant. When the membrane deflects through an angle $d\zeta/dx$, the reforming force in the membrane becomes equal to the transverse component of tension, which gives

$$\frac{d\zeta(x,t)}{dx} = k_s \zeta(x,t), \quad \text{at spring-supported edge.} \tag{7.9}$$

Using the kinematic conditions and eliminating ζ, Eqs. (7.5) and (7.8) yield

$$\frac{\partial \Phi}{\partial y} = 0, \quad \text{at the edges,} \tag{7.10}$$

in the case of a fixed-edge membrane. On the other hand, Eqs. (7.3) and (7.9) yield

$$T \frac{\partial^2 \Phi}{\partial x \partial y} = k_s \frac{\partial \Phi}{\partial y}, \quad \text{at the edges,} \tag{7.11}$$

in the case of a spring-supported flexible membrane. The bottom boundary conditions are given by

$$\frac{\partial \Phi}{\partial y} = 0 \quad \text{on} \quad y = h \quad \text{in the case of finite water depth} \tag{7.12}$$

and

$$\Phi, |\nabla\Phi| \longrightarrow 0 \quad \text{as} \quad y \longrightarrow \infty \quad \text{in the case of infinite water depth.} \tag{7.13}$$

Finally, the far field radiation condition is given by

$$\Phi(x, y, t) = \begin{cases} (I_0 e^{ik_0 x} + R_0 e^{-ik_0 x}) e^{-i\omega t} f_0(y), & \text{as} \quad x \to -\infty, \\ T_0 e^{i(k_0 x - \omega t)} f_0(y), & \text{as} \quad x \to \infty, \end{cases} \tag{7.14}$$

where

$$f_0(y) = \begin{cases} \dfrac{\cosh k_0(h-y)}{\cosh k_0 h}, & \text{in the case of finite water depth,} \\ e^{-k_0 y}, & \text{in the case of infinite water depth,} \end{cases} \tag{7.15}$$

with the wave number k_0 satisfying the dispersion relation in k given by

$$K \equiv \frac{\omega^2}{g} = \begin{cases} k \tanh kh, & \text{in the case of finite water depth,} \\ k, & \text{in the case of infinite water depth.} \end{cases} \tag{7.16}$$

It may be noted that in Eq. (7.14), $I_0 = \dfrac{-Ag}{i\omega}$ is associated with the incident wave of amplitude A assumed to be known and R_0 and T_0 are associated with the amplitude of the reflected and transmitted waves, which are to be determined.

7.3.1 Formulation of the integro-differential equation

Using Green's integral theorem and the free surface Green's function, the basic boundary value problem is converted into a third-order integro-differential equation in terms of the membrane deflection, whose details are discussed next. The method of solution is based on the application of the boundary integral equation approach in the fluid region and suitable expansion of the membrane deflection. From Eq. (7.14), it is clear that the spatial component of the incident wave potential $\phi^i(x, y)$ is given by

$$\phi^i(x, y) = I_0 e^{ik_0 x} f_0(y). \tag{7.17}$$

It is assumed that the spatial velocity potential is written as

$$\phi = \begin{cases} \phi_1 & \text{in the open water region } R_1, \\ \phi_2 & \text{in the membrane-covered region } R_2, \end{cases} \tag{7.18}$$

with

$$\phi_1(x, y) = \phi^i(x, y) + \phi^{sc}(x, y), \tag{7.19}$$

where $\phi^{sc}(x, y)$ is the scattered potential. From Eqs. (7.14) and (7.19), it can be easily derived that

$$\phi^{sc}(x, y) = \begin{cases} R_0 e^{-ik_0 x} f_0(y), & \text{as} \quad x \to -\infty, \\ T_0 e^{ik_0 x} f_0(y), & \text{as} \quad x \to \infty, \end{cases} \tag{7.20}$$

which will determine the unknown constants R_0 and T_0. Once, R_0 and T_0 are obtained, the reflection and transmission coefficients K_r and K_t are obtained from the relation

$$K_r = \left| \frac{R_0}{I_0} \right| \quad \text{and} \quad K_t = \left| \frac{T_0}{I_0} \right|. \tag{7.21}$$

In the boundary integral equation method for wave structure interaction problems to be discussed in the present chapter, Green's identity will be used along with the free surface Green's function which can be obtained from Green's function derived in Chapter 3 for flexural gravity waves (see also Chapter 5). The free surface Green's function $G(x, y; \xi, \tau)$ in the two-dimensional Cartesian coordinate system satisfies the partial differential equation

$$\nabla^2 G = 2\pi \delta(x - \xi)\delta(y - \tau), \tag{7.22}$$

where δ is the well-known Dirac delta function with (ξ, τ) being the source point in the fluid domain. Further, $G(x, y; \xi, \tau)$ satisfies the linearised free surface boundary condition as in Eq. (7.2), the bottom boundary condition and far field condition depending on water depth. The free surface Green's function G in the case of infinite water depth is given by

$$G(x, y; \xi, \tau) = \frac{1}{2\pi} \ln \frac{r}{r'} - \frac{1}{\pi} \int_0^\infty \frac{e^{-k(y+\tau)} \cos k|x - \xi|}{k - K} dk, \tag{7.23}$$

which has the alternate form given by

$$\begin{aligned} G(x, y; \xi, \tau) = {} & -C_0 e^{ik_n |x-\xi| - k_n(y+\tau)} \\ & - \frac{1}{\pi} \int_0^\infty \frac{M(k, \tau) M(k, y) e^{-k|x-\xi|} d\xi}{k \, \Delta(k)}, \end{aligned} \tag{7.24}$$

where $M(k, y) = k \cos ky - K \sin ky$, $r = \sqrt{(x - \xi)^2 + (y - \tau)^2}$, $r' = \sqrt{(x - \xi)^2 + (y + \tau)^2}$ and $C_0 = 2\pi i$. Further, in water of finite depth, the free surface Green's function $G(x, y; \xi, \tau)$ is given by

$$\begin{aligned} G(x, y; \xi, \tau) = {} & \frac{1}{2\pi} \ln \frac{r}{r'} - 2 \int_0^\infty \frac{e^{-kh}}{k} \frac{\sinh ky \sinh k\tau}{\cosh kh} \cos k|x - \xi| dk \\ & - 2 \int_0^\infty \frac{\cosh k(h - y) \cosh k(h - \tau)}{k \sinh kh - K \cosh kh} \frac{\cos k|x - \xi|}{\cosh kh} dk, \end{aligned} \tag{7.25}$$

which has the alternate form as given by

$$G(x, y; \xi, \tau) = \frac{\mathcal{A}_0 \cosh k_0(h - \tau) \cosh k_0(h - y)}{\cosh^2 k_0 h} e^{ik_0|x-\xi|}$$

$$+ \sum_{n=1}^{\infty} \frac{\mathcal{A}_n \cos k_n(h - \tau) \cos k_n(h - y)}{\cos^2 k_n h} e^{-k_n|x-\xi|}, \quad (7.26)$$

where $\mathcal{A}_n$ is given by (which can be derived from Eq. (5.42))

$$\mathcal{A}_n = \begin{cases} \dfrac{4\pi i k_0}{K(1 - Kh) + k_0^2 h}, & \text{for} \quad n = 0, \\[3mm] \dfrac{-4\pi k_n}{K(1 - Kh) - k_n^2 h}, & \text{for} \quad n = 1, 2, ... \end{cases} \quad (7.27)$$

Applying Green's theorem to the velocity potentials ϕ_1 and ϕ_2 as in Eq. (7.18) and the free surface Green's function $G(x, y; \xi, \tau)$, it can be easily derived that

$$2\pi \phi^{sc} = -\int_{\partial R \cup \mathcal{F}} \left(\phi^{sc} \frac{\partial G}{\partial n} - G \frac{\partial \phi^{sc}}{\partial n} \right) dS, \quad \text{for} \quad x \in \mathcal{F} \quad (7.28)$$

$$0 = \int_{\partial R \cup \mathcal{M}} \left(\phi_2 \frac{\partial G}{\partial n} - G \frac{\partial \phi_2}{\partial n} \right) dS \quad \text{for} \quad x \in \mathcal{F}, \quad (7.29)$$

$$0 = -\int_{\partial R \cup \mathcal{F}} \left(\phi^{sc} \frac{\partial G}{\partial n} - G \frac{\partial \phi^{sc}}{\partial n} \right) dS, \quad \text{for} \quad x \in \mathcal{M}, \quad (7.30)$$

$$2\pi \phi_2 = \int_{\partial R \cup \mathcal{M}} \left(\phi_2 \frac{\partial G}{\partial n} - G \frac{\partial \phi_2}{\partial n} \right) dS, \quad \text{for} \quad x \in \mathcal{M}, \quad (7.31)$$

where G is the free surface Green's function as defined in Eqs. (7.22)–(7.26). Using the free surface condition in Eq. (7.2) satisfied by ϕ^{sc} and the Green function G, it is easy to derive that in Eq. (7.30), the integral over $\mathcal{F}$ vanishes. Adding Eqs. (7.30) and (7.31), and using the free surface boundary condition in Eq. (7.2) satisfied by ϕ^{sc} and G, it can be easily derived that

$$2\pi \phi_2 = \int_{\partial R} \left([\phi] \frac{\partial G}{\partial n} - G \left[\frac{\partial \phi}{\partial n} \right] \right) dS + \int_{\mathcal{M}} \left(K \phi_2 - \frac{\partial \phi_2}{\partial \tau} \right) G dS, \text{ for } x \in \mathcal{M},$$

$$(7.32)$$

where $[\phi]$ denotes the difference between the potentials ϕ_2 and ϕ^{sc} and their normal derivatives and is referred as the jumps of the concerned functions. Writing ϕ^{sc} in terms of ϕ_1 as in Eq. (7.19) and using the continuity of the velocity potential and pressure at the interface ∂R, the first integral in Eq. (7.32) yields

$$\int_{\partial R} \left([\phi] \frac{\partial G}{\partial n} - G \left[\frac{\partial \phi}{\partial n} \right] \right) dS = \int_{\partial R} \left(\phi^i \frac{\partial G}{\partial n} - G \frac{\partial \phi^i}{\partial n} \right) dS = 2\pi \phi^i. \quad (7.33)$$

It may be noted that Green's identity is applied to the second integral in

Eq. (7.32) to arrive at the right-hand side result. Using the boundary condition on the membrane-covered mean free surface in Eq. (7.6), Eqs. (7.32) and (7.33) yield the integro-differential equation given by

$$\left(\mu + M\frac{\partial^2}{\partial x^2}\right)\frac{\partial\phi_2}{\partial y} - \frac{K}{2\pi}\int_M G\left(\gamma + M\frac{\partial^2}{\partial\xi^2}\right)\frac{\partial\phi_2}{\partial\tau}dS = -\phi_y^i \quad \text{on} \quad y = 0, \quad (7.34)$$

where $M = T/\rho g$, $\gamma = m\omega^2/\rho g$, $\mu = 1 - \gamma$ and $K = \omega^2/g$. Next, using the kinematic condition in Eq. (7.5) and using the incident potential as in Eq. (7.17), Eq. (7.34) yields

$$\left(\mu + M\frac{\partial^2}{\partial x^2}\right)\eta(x) - \frac{K}{2\pi}\int_M G(x,0;\xi,0)\left(\gamma + M\frac{\partial^2}{\partial\xi^2}\right)\eta(\xi)d\xi = Ae^{ik_0 x}, \quad (7.35)$$

which is an integro-differential equation of the third order in terms of the membrane deflection. Next, the procedure to determine $\eta(x,t)$ will be discussed via two different methods.

7.3.2   System of equations based on wet mode analysis

In this subsection, an eigenfunction expansion method is exploited in terms of the roots of the dispersion relation in the membrane covered region. The cases of water of both infinite and finite depths are discussed separately.

In the case of infinite water depth, from Eq. (7.17), it is clear that the incident wave potential ϕ^i is given by

$$\phi^i = I_0 e^{ik_0 x - k_0 y}. \quad (7.36)$$

Since the flexible membrane is assumed to be of finite length, the membrane deflection is approximated as a linear superposition of the horizontal eigenfunctions associated with the membrane gravity waves in the form given by

$$\eta(x) = \sum_{n=0}^{I}(a_n e^{ip_n x} + b_n e^{-ip_n x}), \quad \text{for} \quad 0 < x < l, \quad (7.37)$$

where a_n and b_n are the unknown amplitudes to be determined and p_n is the root of the membrane-covered gravity wave dispersion relation in p given by

$$(Mp^2 - \gamma + 1)p = K. \quad (7.38)$$

It may be noted that Eq. (7.38) has one real root p_0 and two complex roots p_1 and p_2. However, only one of the complex roots leading to the bounded solution is taken into account in the expansion for $\eta(x)$. Substituting the expression for Green's function G as in Eq. (7.24) without the integral term (which contributes toward the local disturbances) and the membrane deflection $\eta(x)$

as in Eq. (7.37) in the integro-differential equation in Eq. (7.35) and equating the coefficients of $e^{ik_0 x}$ and $e^{-ik_0 x}$, two linear equations are obtained as

$$\sum_{n=0}^{I}(Mp_n^2 + \gamma)k_0 \left(\frac{a_n}{p_n - k_0} - \frac{b_n}{p_n + k_0}\right) - A = 0 \qquad (7.39)$$

and

$$\sum_{n=0}^{I}(Mp_n^2 + \gamma)k_0 \left(\frac{a_n e^{ip_n l}}{p_n + k_0} - \frac{b_n e^{-ip_n l}}{p_n - k_0}\right) = 0. \qquad (7.40)$$

Further, the fixed edge boundary conditions at $x = 0, l$ as in Eq. (7.8) yields

$$\sum_{n=0}^{I}(a_n + b_n) = 0 \qquad (7.41)$$

and

$$\sum_{n=0}^{I}(a_n e^{ip_n l} + b_n e^{-ip_n l}) = 0. \qquad (7.42)$$

In a similar manner, in the case of a flexible membrane having one of the edges at $x = 0$, Eq. (7.8) is satisfied, while in the case of the spring-supported edge condition at $x = l$, Eq. (7.9) is satisfied. Thus, in this case, Eq. (7.37) yields

$$\sum_{n=0}^{I}(ip_n T - k_s)a_n e^{ip_n l} - (ip_n T + k_s)b_n e^{-ip_n l} = 0. \qquad (7.43)$$

Thus, in either case, Eqs. (7.39)–(7.43) yield four equations for the determination of the four unknowns a_n and b_n.

Next, the procedure to determine the reflection and transmission coefficient associated with the wave scattering by the flexible floating membrane is discussed. Adding Eqs. (7.28) and (7.29) and using the free surface boundary condition in Eq. (7.2) satisfied by ϕ^{sc} and G, it can be easily derived that

$$2\pi\phi^{sc} = \int_{\partial R} \left([\phi]\frac{\partial G}{\partial n} - G\left[\frac{\partial \phi}{\partial n}\right]\right)dS + \int_{M}\left(K\phi_2 - \frac{\partial \phi_2}{\partial \tau}\right)GdS, \quad \text{for} \quad x \in \mathcal{F},$$
$$(7.44)$$

where $[\phi]$ denotes the difference between the potentials ϕ_2 and ϕ^{sc} and their normal derivatives and is referred as the jumps of the concerned functions. Proceeding in a similar manner as in case of ϕ_2 in Eq. (7.34) and using the identity in Eq. (7.33) and the membrane-covered free surface condition in Eq. (7.6), it can be easily derived that

$$\phi^{sc} = \phi^i - \frac{1}{2\pi}\int_{M} G(x, y; \xi, \tau)\left(\gamma - M\frac{\partial^2}{\partial \xi^2}\right)\frac{\partial \phi_2}{\partial \tau}dS, \quad \text{for} \quad x \in \mathcal{F}. \qquad (7.45)$$

Substituting for the expression for η from Eq. (7.37), using the kinematic condition in Eq. (7.5) and using the relations in Eq. (7.20) and the definition of reflection and transmission coefficients K_r and K_t in Eq. (7.21), K_r and K_t are obtained in terms of a_n and b_n as given by

$$
\begin{aligned}
K_r &= \sum_{n=0}^{I} \frac{(Mp_n^2 - \gamma)a_n k_0}{(p_n + k_0)}(e^{i(k_0 + p_n)l} - 1) \\
&- \sum_{n=0}^{I} \frac{(Mp_n^2 - \gamma)b_n k_0}{(p_n - k_0)}(e^{i(k_0 - p_n)l} - 1),
\end{aligned}
\tag{7.46}
$$

$$
\begin{aligned}
K_t &= 1 - \sum_{n=0}^{I} \frac{(Mp_n^2 - \gamma)a_n k_0}{(p_n - k_0)}(e^{-i(k_0 - p_n)l} - 1) \\
&+ \sum_{n=0}^{I} \frac{(Mp_n^2 - \gamma)b_n k_0}{(p_n + k_0)}(e^{-i(k_0 + p_n)l} - 1).
\end{aligned}
\tag{7.47}
$$

It may be noted that the asymptotic form of Green's function G as in Eq. (7.24) is used in the derivation of K_r and K_t as discussed above in Eqs. (7.46) and (7.47).

Next, the solution procedure of the integro-differential equation is discussed in the case of finite water depth. In the case of finite water depth, the incident wave potential satisfying Eq. (7.1), the free surface boundary condition (7.2) and the bottom boundary condition (7.12) is given by

$$
\phi^i = -\frac{gA}{i\omega}\frac{\cosh k_0(h - y)}{\cosh k_0 h}e^{ik_0 x},
\tag{7.48}
$$

where k_0 is the same as defined in Eq. (7.16). Proceeding in a similar manner as in the case of infinite water depth, in the case of finite water depth, the membrane deflection is approximated as a linear superposition of the horizontal eigenfunctions associated with the gravity waves in the membrane-covered region in the following form

$$
\eta(x) = \sum_{n=0,I,1}^{\infty} (a_n e^{ip_n x} + b_n e^{-ip_n x}), \quad \text{for} \quad 0 < x < l,
\tag{7.49}
$$

where a_n and b_n are the unknown amplitudes to be determined. Further, $\pm p_n$ are roots of the dispersion relation in p as given by

$$
(\mu + Mp_n^2)p_n \tanh p_n h = K.
\tag{7.50}
$$

The dispersion relation in Eq. (7.50) has one real root, two complex roots and an infinite number of imaginary roots. As in the case of infinite water depth, the complex root having a negative imaginary root is not taken into account. Substituting for the Green's function G as in Eq. (7.26) and the membrane deflection $\eta(x)$ as in Eq. (7.49) and collecting the coefficients of $e^{\pm ik_0 x}$ and

$e^{\pm k_n x}$ as in the case of infinite water depth, from Eq. (7.35) a system of $2(N-1)$ linear equations is derived as

$$\sum_{n=0,I,1}^{N-2} \frac{k_i K(Mp_n^2 - \gamma)}{K(1-Kh)+k_i^2 h}\left(\frac{a_n}{p_n-k_i}-\frac{b_n}{p_n+k_i}\right)+A_i = 0, \qquad (7.51)$$

$$\sum_{n=0,I,1}^{N-2} \frac{k_i K(Mp_n^2 - \gamma)}{K(1-Kh)+k_i^2 h}\left(\frac{-a_n e^{ip_n l}}{p_n+k_i}+\frac{b_n e^{-ip_n l}}{p_n-k_i}\right) = 0, \qquad (7.52)$$

for $i = 0,1,2,...,N-2$ with $A_0 = A$, $A_i = 0$ for $n = 1,2,...,N-2$. It may be noted that the series expansion for $\eta(x)$ is truncated after $N-2$ terms and the series for Green's function is truncated after $N-2$ terms. Further, the edge conditions at the membrane edges as in Eqs. (7.8) and (7.9) yield two equations. Thus, a system of $2N$ equations is obtained for the determination of the $2N$ unknowns a_n and b_n for $n = 0, I, 1, 2, ..., N-2$.

Proceeding in a similar manner as in the case of infinite water depth, the reflection and transmission coefficients in K_r and K_t in terms of a_n and b_n are obtained as

$$K_r = K\sum_{n=0,I,1}^{N-2}(Mp_n^2-\gamma)\left\{\frac{a_n(e^{i(k_0+p_n)l}-1)}{p_n+k_0}-\frac{b_n(e^{i(k_0-p_n)l}-1)}{p_n-k_0}\right\} \qquad (7.53)$$

and

$$K_t = 1 - K\sum_{n=0,I,1}^{N-2}(Mp_n^2-\gamma)\left\{\frac{a_n(e^{-i(k_0-p_n)l}-1)}{p_n-k_0}-\frac{b_n(e^{-i(k_0+p_n)l}-1)}{p_n+k_0}\right\}, \qquad (7.54)$$

where $\mathcal{K} = k_0 K/(K-K^2h+k_0^2 h)$.

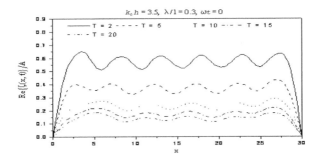

FIGURE 7.2
Membrane deflection having fixed edges.

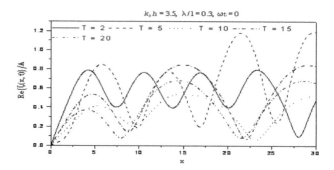

FIGURE 7.3
Membrane deflection having one end fixed and other end a spring-support.

To illustrate the use of the wave diffraction by floating flexible membrane, numerical results are presented in two different cases with $\rho_i = 1025 kg/m^3$, $l = 30m$, $g = 9.8m/s^2$, $d = 0.1cm$, $k_s/\rho g = 5$ and $T = T/\rho g$ varies between 2 to 20 in finite water depth. In Figure 7.2, the membrane deflection is plotted for various values of nondimensional tension parameter T for membranes having both ends fixed with $k_0 h = 3.5$ and $\lambda/l = 0.3$. It is observed that the membrane deflection is zero at the two ends as assumed in the physical problem. Further, the membrane deflection decreases with an increase in the tension parameter. On the other hand, in Figure 7.3, the membrane deflection is plotted for various values of nondimensional tension parameter T for membranes having one end fixed and the other end having spring-support with $k_0 h = 3.5$ and $\lambda/l = 0.3$. It is observed that the membrane deflection is zero at the left end and assumes a finite nonzero value at the spring-supported edge as assumed in the physical model. Further, the membrane deflection decreases with an increase in the tension parameter. From the two figures, it is further clear that the membrane deflection is more in the case of membranes with one of the edges having spring support compared with membranes having both ends fixed.

It may be noted that in the case of infinite water depth, the approximation of the membrane deflection and the free surface Green's function ignore the contribution of the evanescent modes; that is, the local disturbances near the flexible membrane boundary and only the far field effect are taken into consideration. However, in the case of finite water depth, the solution includes the evanescent modes. Thus, the finite depth results will be more accurate compared with the infinite depth results. As a particular case of the finite water depth results, the solution associated with the gravity wave diffraction by a semi-infinite floating flexible membrane can be obtained in a straightforward manner from the solution of the finite plate result by choosing $b_n = 0$. In this case, the solution will demand only one of the edge conditions to be prescribed at $x = 0$. On the other hand, using wide spacing approximation, wave diffraction by single- and multiple-flexible floating membranes can be investigated from the solution of the semi-infinite plate problem and details are deferred here.

7.3.3 System of equations based on dry mode analysis

In the dry mode analysis, the edge condition at $x = 0$ is assumed to be fixed and the other end condition at $x = l$ is assumed to be fixed or spring-supported with spring constant k_s. In order to solve the integro-differential equation in Eq. (7.35), the mode expansion method is used as discussed below.

Assuming the membrane motion is simple harmonic in time with angular frequency ω, the membrane deflection can be written in the form $\zeta(x, t) = \eta(x)e^{-i\omega t}$. Assuming that the mass is negligible, the equation of the membrane with zero-forcing term yields

$$\frac{d^2\eta}{dx^2} + \lambda^2\eta = 0, \tag{7.55}$$

where $\lambda^2 = m\omega^2/T$. Under the assumption that both ends of the membrane are fixed, Eqs. (7.8) and (7.55) yield

$$\eta(x) = \sum_{n=1}^{\infty} a_n\psi_n(x), \tag{7.56}$$

where a_n is a constant to be determined, and ψ_n is given by

$$\psi_n(x) = \frac{2}{l}\sin\lambda_n x, \tag{7.57}$$

where λ_n is given by $\lambda_n = n\pi/l$. On the other hand, in the case of a membrane which is fixed at the end $x = 0$ as in Eq. (7.8) and has the spring-supported edge condition as in Eq. (7.9) at $x = l$, λ_n satisfies

$$\tan\lambda_n l = T\lambda_n/k_s, \quad n = 1, 2, \dots. \tag{7.58}$$

The expansion of $\eta(x)$ in terms of $\psi_n(x)$ is the modal function expansion method or mode expansion method. It may be noted that $\psi_n(x)$ is an orthonormal function. In the context of the present chapter, $\psi_n(x)$ is referred as a dry mode. Using the orthonormal characteristics of $\psi_n(y)$ and truncating the series after N terms, it can be easily derived that

$$\left\{\mu - M\lambda_m^2\right\}a_m + \frac{K}{2\pi}\sum_{n=1}^{N}\int_0^l\int_0^l\left(\gamma + M\lambda_n^2\right)a_n\psi_m(x)\psi_n(\xi)Gdxd\xi$$

$$= A\int_0^l e^{ik_0x}\psi_m(x)dx, \quad \text{for} \quad m = 1, 2, \dots, N. \tag{7.59}$$

On simplification, Eq. (7.59) yields a system of equations for the unknown coefficients a_n given by

$$\left\{\mu - M\lambda_m^2\right\}a_m + \sum_{n=1}^{N}\left(\gamma + M\lambda_n^2\right)g_{mn}a_n = c_m, \quad \text{for} \quad m = 1, 2, ..., N, \quad (7.60)$$

where

$$g_{mn} = \frac{k_0}{2\pi}\int_0^l\int_0^l G(x, 0, \xi, 0)\psi_m(x)\psi_n(\xi)dxd\xi, \quad c_m = A\int_0^l e^{ik_0 x}\psi_m(x)dx.$$

Proceeding in a similar manner as in the case of wet mode analysis discussed in the previous subsection, the reflection and transmission coefficients K_r and K_t defined in Eq. (7.21) are obtained in terms of a_n and b_n as given by

$$K_r = \sum_{n=1}^{N}(M\lambda_n^2 - \gamma)k_0\left\{\frac{a_n(e^{i(k_0+\lambda_n)l} - 1)}{(\lambda_n + k_0)} - \frac{b_n(e^{i(k_0-\lambda_n)l} - 1)}{(p_n - k_0)}\right\}, \quad (7.61)$$

$$K_t = 1 - \sum_{n=1}^{N}(M\lambda_n^2 - \gamma)k_0\left\{\frac{a_n(e^{-i(k_0-\lambda_n)l} - 1)}{(\lambda_n - k_0)} + \frac{b_n(e^{-i(k_0+p_n)l} - 1)}{(p_n + k_0)}\right\}, \quad (7.62)$$

in the case of infinite water depth. Further, in the case of finite water depth, the reflection and transmission coefficients are obtained as

$$K_r = \mathcal{K}\sum_{n=1}^{N}(M\lambda_n^2 - \gamma)\left\{\frac{a_n(e^{i(k_0+\lambda_n)l} - 1)}{\lambda_n + k_0} - \frac{b_n(e^{i(k_0-\lambda_n)l} - 1)}{\lambda_n - k_0}\right\} \quad (7.63)$$

and

$$K_t = 1 - \mathcal{K}\sum_{n=1}^{N}(M\lambda_n^2 - \gamma)\left\{\frac{a_n(e^{-i(k_0-\lambda_n)l} - 1)}{\lambda_n - k_0} - \frac{b_n(e^{-i(k_0+\lambda_n)l} - 1)}{\lambda_n + k_0}\right\}, \quad (7.64)$$

where $\mathcal{K} = k_0 K/(K - K^2 h + k_0^2 h)$. It may be noted that the reflection and transmission coefficients defined above satisfy the energy relation

$$K_r^2 + K_t^2 = 1, \quad (7.65)$$

which is used to check the accuracy of the computational results.

7.4   Conclusion

The role of the boundary integral equation method in wave structure interaction problems is demonstrated by analysing the diffraction of surface waves by a floating flexible membrane in the Cartesian coordinate system. In terms of the vertical deflection of the membrane, the boundary value problem is converted into an integro-differential equation whose solution is obtained via two different methods. The first method is referred to as the wet mode analysis. In this method, the membrane deflection is approximated by an expansion in terms of the horizontal eigenfunctions involving the eigenfunctions which are the roots of the membrane-covered gravity wave dispersion relation. The second method is referred to as the dry mode expansion method in which the membrane deflection is expanded in terms of its mode functions and is fitted into the integro-differential equation to determine a system of equations from which the required physical quantities are derived. The results can be suitably generalised to deal with wave interaction with two-dimensional flexible floating membranes and circular flexible floating membranes satisfying various types of edge conditions in the cases of water of both finite and infinite depths. The method discussed can be easily generalised to deal with an oblique gravity wave interaction with flexible floating membranes. One of the important aspects of these classes of problems is to study the mathematical convergence of various types of wave structure interaction problems analysed based on the boundary integral equation method. Apart from appropriately choosing the membrane deflection, the integro-differential equation can be attempted for solution directly. Further, the effect of uniform current on wave diffraction by floating membrane can be studied using a similar approach as in [48], and the effect of porosity in the floating structure can be included. Some of these problems are highlighted as exercises at the end of the chapter.

7.5   Examples and exercises

Exercise 7.1 Discuss the diffraction of surface water waves by a flexible floating membrane in two dimensions in the presence of current in the cases of water of both finite and infinite water depths using the boundary integral equation method.

Exercise 7.2 Discuss the diffraction of surface water waves by a flexible submerged membrane in two dimensions in the presence of current in the cases of water of both finite and infinite water depths using the boundary integral equation method.

Exercise 7.3 Discuss the surface wave interaction with a flexible floating/submerged membrane in two dimensions kept near a sea wall in the cases of water of both finite and infinite water depths using the boundary integral equation method.

Exercise 7.4 Discuss the diffraction of oblique surface water waves by a two-dimensional flexible floating permeable membrane in the cases of water of both finite and infinite water depths using the boundary integral equation method.

Exercise 7.5 Discuss the oblique water wave diffraction by a flexible floating membrane in two dimensions in the presence of current in the cases of water of both finite and infinite water depths based on the boundary integral equation method.

Exercise 7.6 Discuss the diffraction of a plane progressive wave by a circular flexible floating membrane in the cases of water of both finite and infinite water depths based on the boundary integral equation method.

Exercise 7.7 Discuss the effect of variation in bottom topography on wave diffraction by a finite floating membrane in the cases of water of both finite and infinite water depths based on the boundary integral equation method.

Exercise 7.8 Discuss the diffraction of a finite number of flexible membranes each of finite length kept in water of both finite and infinite water depths based on the boundary integral equation method. The solution of wave diffraction by multiple floating membranes can be obtained by the wide spacing approximation method discussed in Chapter 6 from the solution of a single finite membrane and the results can be compared for the accuracy of the solution.

References

[1] A. G. Abul-Azm. Wave difffraction by double flexible breakwaters. *Appl. Ocean Res.*, 16:87–99, 1994.

[2] A. I. Andrianov and A. J. Hermans. The influence of water depth on the hydroelastic response of a very large floating platform. *Marine Structure*, 16:355–371, 2003.

[3] A. I. Andrianov and A. J. Hermans. Hydroelasticity of a circular plate on water of finite and infinite depth. *J. of Fluids and Structures*, 20:719–733, 2005.

[4] A. I. Andrianov and A. J. Hermans. Hydroelastic behavior of a floating ring-shaped plate. *J. of Eng. Math.*, 54:31–48, 2006.

[5] N. J. Balmforth and R. V. Craster. Ocean waves and ice sheets. *J. Fluid. Mech.*, 395:89–124, 1999.

[6] K. A. Belibassakis and G. A. Athanassoulis. A coupled-mode technique for weakly nonlinear wave interaction with large floating structures lying over variable bathymetry regions. *Appl. Ocean Res.*, 28:59–76, 2006.

[7] L. G. Bennetts, N. R. T. Biggs, and D. Porter. A multi-mode approximation to wave scattering by ice sheets of varying thickness. *J. Fluid Mech.*, 579:413–443, 2007.

[8] L. G. Bennetts and V. A. Squire. Linear wave forcing of an array of axisymmetric ice floes. *IMA J. Appl. Math*, 75:108–138, 2010.

[9] J. Bhattacharjee, D. Karmakar, and T. Sahoo. Transformation of flexural gravity waves by heterogeneous boundaries. *J. Eng. Math.*, 62:173–188, 2007.

[10] J. Bhattacharjee and T. Sahoo. Flexural gravity wave problems in two-layer fluids. *Wave Motion*, 45(11):133–153, 2008.

[11] J. Bhattacharjee and T. Sahoo. Generation of flexural gravity waves in the presence of uniform current. *J. Marine Science and Technology*, 13(2):138–146, 2008.

[12] J. Bhattacharjee and T. Sahoo. Interaction of flexural gravity waves with jet like current in shallow water. *Ocean Engineering*, 36(11):831–841, 2009.

[13] R. E. D. Bishop and W. G. Price. Hydroelasticity of ships. *Cambridge University Press*, 26:245–256, 1979.

[14] A. Chakrabarti. On the solution of the problem of scattering of surface-water waves by the edge of an ice cover. *Proc. R. Soc. Lond. A*, 456:1087–1099, 2000.

[15] A. Chakrabarti and Hamsapriye. On some general hybrid transforms. *J. Comput. Appl. Math.*, 116(1):157–165, 2000.

[16] X. Chen, Y. Wu, W. Cui, and J. J. Jensen. Review of hydroelasticity theories for global response of marine structures. *Ocean Engineering*, 33:245–256, 2006.

[17] I. H. Cho, S. T. Kee, and M.H. Kim. The performance of flexible-membrane wave barriers in oblique incident waves. *Applied Ocean Res.*, 19:171–182, 1997.

[18] I. H. Cho and M. H. Kim. Wave absorbing system using inclined perforated plates. *J. Fluid. Mech.*, 608:1–20, 2008.

[19] I. H. Cho and M.H. Kim. Interactions of a horizontal flexible membrane with oblique waves. *J. Fluid Mech.*, 367:139–161, 1998.

[20] I. H. Cho and M.H. Kim. Wave deformation by a submerged flexible circular disk. *Applied Ocean Res.*, 21(5):263–280, 1999.

[21] H. Chung and C. Fox. Propagation of flexural waves at the interface between floating plates. *Intl. J. of Offshore and Polar Eng.*, 12(3):163–170, 2002.

[22] A. T. Chwang. A porous wave-maker theory. *J. Fluid Mech.*, 132:395–406, 1983.

[23] A. T. Chwang and Z. N. Dong. Wave trapping due to porous plate. *Proc. of the 15th ONR Symposium Naval Hydrodynamics*, pp. 407–414, 1984.

[24] E. T. Copson. *Asymptotic expansions*. Cambridge, U.K., Cambridge University Press, 1st edition, 1965.

[25] R. A. Dalrymple and P. A. Martin. Wave diffraction through offshore breakwaters. *J. Waterway, Port, Coastal and Ocean Eng.*, 116(6):727–741, 1990.

[26] D. Das and B. N. Mandal. Construction of wave-free potential in the linearized theory of water waves. *J. Marine Sci. Appl.*, 9:347–354, 2010.

[27] J. W. Davys, R. J. Hosking, and A. D. Sneyd. Waves due to a steadily moving source on a floating ice plate. *J. Fluid Mech.*, 158:269–287, 1985.

[28] R. G. Dean and R. A. Dalrymple. *Water waves Mechanics for Engineers and Scientists*. Singapore, World Scientific, 1991.

[29] M. W. Dingemans. *Water wave propagation over uneven bottoms, Part-I — Linear wave propagation, Advanced series on ocean engineering — Vol. 13*. Singapore, World Scientific, 1997.

[30] P. A. M. Dirac. *The principles of quantum mechanics*. Oxford, U.K., Oxford University Press, 2nd edition, 1935.

[31] D. G. Duffy. *Green's functions with applications*. New York, Chapman & Hall/CRC, 2001.

[32] D. V. Evans. The wide-spacing approximation applied to multiple scattering and sloshing problems. *J. Fluid Mech.*, 210:647–658, 1990.

[33] D. V. Evans and R. Porter. Wave scattering by narrow cracks in ice sheets floating on water of finite depth. *J. Fluid Mech*, 484:143–165, 2003.

[34] J. Falnes. *Ocean waves and oscillating systems*. Cambridge, U.K., Cambridge University Press, 1st edition, 2002.

[35] A. B. Finkelstein. The initial value probelm for transient water waves. *Comm. Pure Appl. Math.*, 10:511–522, 1957.

[36] C. J. Fitzgerald and M. H. Meylan. Generalized eigenfunction method for floating bodies. *J. Fluid Mech*, 667:544–554, 2011.

[37] C. Fox and V. A. Squire. Coupling between the ocean and an ice shelf. *Annals of Glaciology*, 15:101–108, 1991.

[38] C. Fox and V. A. Squire. On the oblique reflection and transmission of ocean waves at shore fast sea ice. *Phil. Trans. R. Soc. Lond. A*, 347:185–218, 1994.

[39] R. P. Gao, Z. Y. Tay, C. M. Wang, and C. G. Koh. Hydroelastic response of very large floating structure with a flexible line connection. *Ocean. Eng.*, 38:1957–1966, 2011.

[40] R. Gayen and B. N. Mandal. Motion due to fundamental singularities in finite depth water with an elastic solid cover. *Fluid Dyn. Research*, 38:224–240, 2006.

[41] M. D. Greenberg. *Application of Green's Functions in Science and Engineering*. Englewood Cliffs, NJ, Prentice Hall, 1971.

[42] A. G. Greenhill. Wave motion in hydrodyanamics. *Am. J. Math*, 9:62–112, 1887.

[43] M. Hassan, M. H. Meylan, and M. A. Peter. Water wave scattering by submerged elastic plates. *Quarterly J. of Mechanics and Applied Math.*, 62(3):321–344, 2009.

[44] T. H. Havelock. Forced surface waves on water. *Philos. Mag.*, 8:569–578, 1929.

[45] G.-H. He and M. Kashiwagi. Nonlinear analysis on wave-plate interaction due to disturbed vertical elastic plate. *J. Hydrodynamics, Ser. B*, 22(5):507–512, 2010.

[46] G. M. Hegarty and V. A. Squire. A boundary-integral method for the interaction of large-amplitude ocean waves with a compliant floating raft such as a sea-ice floe. *J. Eng. Math.*, 62:355–372, 2008.

[47] S. R. Heller. *Hydroelasticity: Advances in hydrosciences 1.* New York, Academic Press, 1964.

[48] A. J. Hermans. A boundary element method for the interaction of free-surface waves with a very large flexible platform. *J. Fluids and Structures*, 14:943–956, 2000.

[49] A. J. Hermans. Interaction of free-surface waves with floating flexible strips. *J. Eng. Math*, 49:133–147, 2004.

[50] A. J. Hermans. Free-surface wave interaction with a thick flexible dock or very large floating platform. *J. of Eng. Math.*, 58:77–90, 2007.

[51] A. J. Hermans. *Water waves and ship hydrodynamics: An introduction.* New York, Springer, 2nd edition, 2011.

[52] S.E. Hirdaris and P. Tamarel. Hydroelasticity of ships: Recent advances and future trends. *J. Eng. Maritime Environment*, 223(3):305–330, 2009.

[53] D. C. Hong and S. Y. Hong. Hydroelasticresponses and drift forces of a very-long floating structure equipped with a pin-connected oscillating-water-column breakwater system. *Ocean. Eng.*, 34:696–708, 2007.

[54] D. C. Hong, S. Y. Hong, and S. W. Hong. Reduction of hydroelastic responses of a very-long floating structure by a floating oscillating-water-column breakwater system. *Ocean. Eng.*, 33:610–634, 2006.

[55] T. Ikoma and K. Masuda. Hydroelastic behaviors of VLFS supported by many aircushions with the three-Dimensional linear theory. *J. Offshore Mech. and Arctic Engrg.*, 134:011104–011108, 2012.

[56] R. S. Johnson. *A modern introduction to the theory of water waves.* Cambridge, U.K., Cambridge University Press, 1st edition, 1997.

[57] I. G. Jonsson and J. D. Wang. Current-depth refraction of water waves. *Ocean Eng.*, 62(7):153–171, 1980.

[58] R. P. Kanwal. *Linear integral equations theory and techniques*. London, UK, Academic Press, 1st edition, 1971, Boston, Birkhauser, 2nd edition, 1996.

[59] D. Karmakar, J. Bhattacharjee, and T. Sahoo. On the hydroelastic behaviour of very large floating structures to gravity waves. *Intl. J. Eng. Sc.*, 45(10):807–828, 2007.

[60] D. Karmakar, J. Bhattacharjee, and T. Sahoo. Wave interaction with multiple articulated floating elastic plates. *J. Fluids Struct.*, 25:1065–1078, 2009.

[61] D. Karmakar, J. Bhattacharjee, and T. Sahoo. Oblique flexural gravity-wave scattering due to changes in bottom topography. *J. Eng. Math.*, 66:325–341, 2010.

[62] D. Karmakar and T. Sahoo. Scattering of waves by articulated floating elastic plate in water of infinite depth. *Marine Struct.*, 18:451–471, 2005.

[63] D. Karmakar and T. Sahoo. Gravity wave interaction with floating membrane with abrupt change in bottom topography. *Ocean Engrg.*, 35(7):598–615, 2008.

[64] D. Karmakar and C. Guedes Soares. Scattering of gravity waves by a moored finite floating elastic plate. *Appl. Ocean Res.*, 34:135–149, 2012.

[65] M. Kashiwagi. A time-domain mode-expansion method for calculating transient elastic responses of a pontoon-type VLFS. *J. Mar. Sci. Technol.*, 5:89–100, 2000.

[66] M. Kashiwagi. A time-domain mode-expansion method for calculating transient elastic responses of a pontoon-type VLFS. *J. Mar. Sci. and Technol.*, 5:89–100, 2000.

[67] M. Kashiwagi. Research on hydroelastic responses of VLFS: Recent progress and future work. *Intl. J. Offshore Polar Eng.*, 10:81–90, 2000.

[68] A. D. Kerr. The critical velocities of a load moving on a floating ice plate that is subjected to in-plane forces. *Cold Regions Sci. and Technol.*, 6:267–274, 1983.

[69] T. I. Khabakhpasheva and A. A. Korobkin. Hydroelastic behaviour of compound floating plate in waves. *J. Eng. Math.*, 44:21–40, 2002.

[70] M. H. Kim and S. T. Kee. Flexible membrane wave barrier I: Analytic and numerical solutions. *J. Waterway Port Coastal and Oc. Eng.*, 122(1):46–53, 1996.

[71] A. Korobkin, E. I. Parau, and J. M. V. Broeck. The mathematical challenges and modelling of hydroelasticity. *Phil. Trans. R. Soc. A*, 369:2803–2812, 2011.

[72] G.A. Kriegsmann. Scattering matrix analysis of a photonic Fabry-Perot resonator. *Wave Motion*, 37:43–61, 2003.

[73] P. S. Kumar and T. Sahoo. Wave interaction with a flexible porous breakwater in a two-layer fluid. *J. Eng. Mech.*, 132(9):1007–1014, 2006.

[74] J. B. Lawrie. On eigenfunction expansions associated with wave propagation along ducts with wavebearing boundaries. *IMA J. Appl. Math.*, 72:376–394, 2007.

[75] J. B. Lawrie. Orthogonality relations for fluid-structural waves in a three-dimensional, rectangular duct with flexible walls. *Proc. R. Soc. A*, 465:2347–2367, 2009.

[76] J. B. Lawrie. Comments on a class of orthogonality relations relevant to fluid-structure interaction. *Meccanica*, 47(3):783–788, 2012.

[77] J. B. Lawrie and I. D. Abrahams. An orthogonality relation for a class of problems with high-order boundary conditions; applications in sound-structure interaction. *Q. J. Mech. Appl. Math.*, 52(2):161–181, 1999.

[78] J. B. Lawrie and I. D. Abrahams. A brief historical perspective of the Wiener-Hopf technique. *J. Eng. Math.*, 59:351–358, 2007.

[79] W. K. Lee and E. Y. M. Lo. Surface-penetrating flexible membrane wave barriers of finite draft. *Ocean Eng.*, 29(14):1781–1804, 2002.

[80] M. J. Lighthill. *Introduction to Fourier analysis and generalised functions.* Cambridge, UK, Cambridge University Press, 1st edition, 1959.

[81] C. M. Linton and M. McIver. The interaction of waves with horizontal cylinders in two-layer fluids. *J. Fluid Mech.*, 304:213–229, 1995.

[82] C. M. Linton and P. McIver. *Handbook of mathematical techniques for wave/structure interactions.* Boca Raton, FL, Chapman Hall/CRC, 2001.

[83] D. Q. Lu and S. Q. Dai. Flexural- and capillary-gravity waves due to fundamental singularities in an inviscid fluid of finite depth. *Int. J. Eng. Sci.*, 46:1183–1193, 2008.

[84] A. G. Mackie. Initial value problem in water wave theory. *J. Aust. Math. Soc.*, 3:340–350, 1963.

[85] E. B. Magrab. *Vibrations of elastic structural members.* Sijthoff & Noordhoff. International Publishers B. V., Alphen aan den Rijn, The Netherlands. 4th edition, 1979.

[86] P. Maiti and B. N. Mandal. Water waves generated due to initial axisymmetric disturbances in water with an ice cover. *Arch. Appl. Mech.*, 74:629–636, 2005.

[87] S. R. Manam, J. Bhattacharjee, and T. Sahoo. Expansion formulae in wave structure interaction problems. *Proc. Roy. Soc. Lond. Ser. A*, 462(2065):263–287, 2006.

[88] S. R. Manam and T. Sahoo. Wave past porous structures in two-layer fluid. *J. Eng. Math.*, 54(4):355–377, 2005.

[89] B. N. Mandal and A. Chakrabarti. A generalisation to the hybrid Fourier transform and its application. *Appl. Math. Letters*, 16:703–708, 2003.

[90] B. N. Mandal and A. Chakrabarti. *Applied singular integral equations*. Boca Raton, FL, CRC Press Taylor & Francis Group, 2011.

[91] A. V. Marchenko. Surface wave diffraction at a crack in sheet ice. *Fluid Dynamics*, 28(2):230–237, 1993.

[92] P. McIver. Wave forces on adjacent floating bridges. *Appl. Ocean Res.*, 8:67–75, 1986.

[93] C. C. Mei. *The applied dynamics of ocean surface waves*. Singapore, World Scientific, 1st edition, 1989.

[94] C. C. Mei. *Mathematical analysis in engineering*. New York, Cambridge University Press, 1st edition, 1995.

[95] M. H. Meylan. Spectral solution of time-dependent shallow water hydroelasticity. *J. Fluid Mech.*, 454:387–402, 2002.

[96] M. H. Meylan and I. V. Sturova. Time-dependent motion of a two-dimensional floating elastic plate. *J. Fluids Struct.*, 25:445–460, 2009.

[97] J. Miles and A. D. Sneyd. The response of a floating ice sheet to an accelerating line load. *J. Fluid Mech.*, 497:435–439, 2003.

[98] F. Milinazzo, M. Shinbrot, and N. W. Evans. A mathematical analysis of the steady response of floating ice to the uniform motion of a rectangular load. *J. Fluid Mech.*, 287:173–197, 1995.

[99] L. M. Milne-Thomson. *Theoretical hydrodynamics*. New York, Dover, 6th edition, 1996.

[100] S. C. Mohapatra and T. Sahoo. On capillary gravity wave motion in two layer fluids. *J. Eng. Math.*, 71(3):253–277, 2011.

[101] S. C. Mohapatra and T. Sahoo. Surface gravity wave interaction with elastic bottom. *Appl. Ocean Res.*, 33(1):31–40, 2011.

[102] S. C. Mohapatra, R. Ghoshal, and T. Sahoo. Effect of compression on wave differaction by a floating elastic plate. *J. Fluids and Structures*, DOI: 10.1016/7.jfluidstructs.2012.07.005, 2012.

[103] R. Mondal, S. K. Mohanty, and T. Sahoo. Expansion formulae for wave structure interaction problems in three dimensions. *IMA J. Appl. Math.*, doi:10.1093/imamat/hxr044, 2011.

[104] R. Mondal and T. Sahoo. Wave structure interaction problems for two-layer fluids in three dimensions. *Wave Motion*, 49:501–524, 2012.

[105] F. Montiel, L. G. Bennetts, and V. A. Squire. The transient response of floating elastic plates to wavemaker forcing in two dimensions. *J. Fluid Struct.*, 28:416–433, 2012.

[106] Y. Namba and M. Ohkusu. Hydroelastic behavior of floating artificial islands in waves. *Intl. J. Offshore and Polar Eng.*, 9:39–47, 1999.

[107] J. N. Newman. Wave effects on deformable bodies. *Appl. Ocean Res.*, 16(1):47–59, 1994.

[108] T. C. Nguyen and R. W. Yeung. Unsteady three-dimensional sources for a two-layer fluid of finite depth and their applications. *J. Eng. Math.*, 70:67–91, 2011.

[109] M. Ohkusu and Y. Namba. Hydroelastic analysis of a large floating structure. *J. Fluids and Structures*, 19:543–555, 2004.

[110] S. Ohmatsu. Overview: Research on wave loading and responses of VLFS. *Marine Struct.*, 18:149–168, 2005.

[111] E. I. Parau and J.-M. Vanden-Broeck. Three-dimensional waves beneath an ice sheet due to a steadily moving pressure. *Phil. Trans. R. Soc. A*, 369(1947):2973–2988, 2011.

[112] D. H. Peregrine. Interaction of water waves and currents. *Adv. Appl. Mech.*, (16):9–117, 1976.

[113] D. Porter and R. Porter. Approximations to wave scattering by an ice sheet of variable thickness over undulating bed topography. *J. Fluid Mech.*, 509:145–179, 2004.

[114] R. Porter and D.V. Evans. Scattering of flexural waves by multiple narrow cracks in ice sheets floating on water. *Wave Motion*, 43:425–443, 2006.

[115] X. Ren and K.-H. Wang. Mooring lines connected to floating porous breakwaters. *Intl. J. of Eng. Sciences*, 32(10):1511–1530, 1994.

[116] P. F. Rhodes-Robinson. On the generation of water waves at an inertial surface. *J. Austral. Math., Soc. Ser. B*, 25:366–383, 1984.

[117] P. F. Rhodes-Robinson. Fundamental singularities in the theory of water waves with surface tension. *Bull. Austral. Math. Soc.*, 2:317–333, 1970.

[118] H. R. Riggs, K. M. Niimi, and L. L. Huang. Two benchmark problems for three-dimensional, linear hydroelasticity. *J. Offshore Mech. Arctic Eng.*, 129:149–157, 2007.

[119] M. Riyansyah, C. M. Wang, and Y. S. Choo. Connection design for two-floating beam system for minimum hydroelastic response. *Marine Structures*, 23:67–87, 2010.

[120] T. Sahoo, M. M. Lee, and A. T. Chwang. Trapping and generation of water waves by vertical porous structures. *J. Eng. Mech.*, 126:1074–1082, 2000.

[121] T. Sahoo, T. L. Yip, and A. T. Chwang. Scattering of surface waves by a semi-infinite floating elastic plate. *Physics of Fluids*, 13(11):3215–3222, 2001.

[122] S. H. Schot. Eighty years of Sommerfeld's radiation condition. *Historia Math.*, 19(4):385–401, 1992.

[123] R. M. S. M. Schulkes, R. J. Hosking, and A. D. Sneyd. Waves due to a steadily moving source on a floating ice plate. Part 2. *J. Fluid Mech.*, 180:297–318, 1987.

[124] V. A. Squire. Of ocean waves and sea-ice revisited. *Cold Reg. Sc. Tech.*, 49(2):110–133, 2007.

[125] V. A. Squire. Synergies between VLFS hydroelasticity and sea-ice research. *Intl. J. of Offshore and Polar Eng.*, 18(3):1–13, 2008.

[126] V. A. Squire. Past, present and impendent hydroelastic challenges in the polar and subpolar seas. *Phil. Trans. R. Soc. A*, 369:2813–2831, 2011.

[127] V. A. Squire, R. J. Hosking, A. D. Kerr, and P. J. Langhorne. *Moving loads on ice plates*. Dordrecht, The Netherlands, Kluwar Academic Publishers, 1996.

[128] V. A. Squire and G. M. Hegarty. A boundary-integral method for the interaction of large-amplitude ocean waves with a compliant floating raft such as a sea-ice floe. *J. Eng. Math.*, 62:355–372, 2008.

[129] J. J. Stoker. *Water waves: The mathematical theory with applications*. New York, Interscience, 2nd edition, 1992.

[130] I. V. Sturova. The oblique incidence of surface waves onto the elastic band. *Proc. 2nd Intl. Conf. on Hydroelasticity in Marine Tech., Fukuoka, Japan*, pp. 239–245, 1998.

[131] I. V. Sturova. Unsteady behavior of an elastic beam floating on shallow water under external loading. *J. Appl. Mech. Tech. Physics*, 43(3):415–423, 2002.

[132] I. V. Sturova. Unsteady behaviour of an elastic beam floating on the surface of an infinitely deep fluid. *J. Appl. Mech. and Tech. Physics*, 47:71–78, 2006.

[133] I. V. Sturova. Unsteady behaviour of heterogeneous elastic beam floating on shallow water. *J. Appl. Math. Mech.*, 72:704–714, 2008.

[134] I. V. Sturova. Time-dependent response of a heterogeneous elastic plate floating on shallow water of variable depth. *J. Fluid Mech.*, 637:305–325, 2009.

[135] I. V. Sturova. Unsteady behavior of an elastic articulated beam floating on shallow water. *J. Appl. Mech. Tech. Physics*, 50(4):589–598, 2009.

[136] P. Suresh Kumar, S. R. Manam, and T. Sahoo. Wave scattering by flexible porous membrane barrier in a two-layer fluid. *J. Fluids and Structures*, 23:633–647, 2007.

[137] R. C. Thorne. Multipole expansions in the theory of surface waves. *Math. Proc. Camb. Phil. Soc.*, 49(4):707–716, 1953.

[138] C. D. Wang and M. H. Meylan. The linear wave response of a floating thin plate on water of variable depth. *Appl. Ocean Res.*, 24(3):163–174, 2002.

[139] C. M. Wang, Z. Y. Tay1, K. Takagi, and T. Utsunomiya. Literature review of methods for mitigating hydroelastic response of VLFS under wave action. *Appl. Mech. Review*, 63:030802–1–030802–18, 2010.

[140] K.-H. Wang and X. Ren. An effective wave trapping system. *Ocean Eng.*, 21(2):155–178, 1994.

[141] E. Watanabe, T. Utsunomiya, and C. M. Wang. Hydroelastic analysis of pontoontype VLFS: A literature survey. *Eng. Struct.*, 26:245–256, 2004.

[142] J. V. Wehausen and E. V. Laitone. *Surface waves in encyclopaedia of physics*, Vol. IX. Berlin, Springer Verlag, 1960.

[143] A. N. Williams. Dual floating breakwater. *Ocean Eng.*, 20(3):215–232,, 1993.

[144] A. N. Williams, P. T. Greiger, and W. G. McDougal. Flexible floating breakwater. *J. Waterways, Port, Coastal and Ocean Eng.*, 117(5):429–450, 1991.

[145] A. N. Williams and K. H. Wang. Flexible porous wave barrier for enhanced wetlands habitat restoration. *J. Eng. Mech.*, 129(1):1–8, 2003.

[146] T. D. Williams and R. Porter. The effect of submergence on the scattering by the interface between two semi-infinite sheets. *J. Fluids Struct.*, 25(5):777–793, 2009.

[147] T. D. Williams and V. A. Squire. Wave propagation across an oblique crack in an ice sheet. *Intl. J. of Offshore and Polar Eng.*, 12(3):157–162, 2002.

[148] W. E. Williams. *Partial differential equations*. Oxford, Clarendon Press, 1980.

[149] Y. S. Wu and W. C. Cui. Advances in the three-dimensional hydroelasticity of ships. *Part M: J. Eng. Maritime Environment*, 223(3):331–348, 2009.

[150] D. Xia, J. W. Kim, and R. C. Ertekin. On the hydroelastic behavior of two-dimensional articulated plates. *Marine Struct.*, 13:261–278, 2000.

[151] X. Xia and H. T. Shen. Nonlinear interaction of ice cover with shallow water waves in channels. *J. Fluid Mech.*, 467:259–268, 2002.

[152] T. L. Yip, T. Sahoo, and A. T. Chwang. Trapping of surface waves by porous flexible structures. *Wave Motion*, 35:41–54, 2002.

[153] X. Yu. Diffraction of water waves by porous breakwater. *J. Waterway, Port, Coastal, and Ocean Eng.*, 121:275–282, 1995.

[154] C.-Y. Yueh and S.-H. Chuang. A boundary element model for a partially piston-type porous wave energy converter in gravity waves. *Eng. Anal. with Boundary Elements*, 36:658–664, 2012.

[155] F. Zhao, W. Bao, T. Kinoshita, and H. Itakura. Theoretical and experimental study on a porous cylinder floating in waves. *J. of Offshore Mech. and Arctic Eng.*, 133(1), 2011.

Index

A

Articulated submerged plates, wave scattering caused by. *See under* Scattering

B

Bending moment, 103
Bernoulli's equations, 16, 17, 26, 120
    linearised, 174
Boundary integral equation method, 1
Boundary value problems (BVP), 20–21
    expansion formulae in finite water depth, 48–49
    expansion formulae in infinite water depth, 45, 46–47
    Green's function technique for ordinary differential equations (ODEs), 80–82
    Laplace equations, associated with, 66–67
    overview of solutions, 39
    wave interaction with rigid structures, 43
    wave structure interaction problems, two-dimensional, 43–45
Breaking, waves, 22
Breakwaters, 117, 120

C

Cauchy residue theorem, 13, 47, 89
Complex function theory, 1
Compressive force, 65

Conservation of energy flux, law of, 180
Continuum mechanics, 9–10
Convolution formula, 153
Cordgrass seedlings, 194
Currents, ocean, 6

D

D'Alembert solution, 13
Deep water, wave dispersion in, 20
Deflection, slope of, 102, 103
Delta distribution, 80
Delta function, 79–80
Depth, water
    deep water, 20
    finite, 21, 42, 89–91, 153, 156
    infinite, 21, 45, 46–47, 153, 156
    shallow water waves, 21
Diffraction, wave
    dual membranes, caused by, 194
    Green's function, use of; *see under* Green's function
    integro-differential equations, using, 197–200
    mathematical formulation, 195–197
    membranes, flexible, 200, 202
    overview, 193–194
    study of, 193
    surface gravity waves, of, 194
Dirac delta function, 46, 51, 80, 161
Dirac, P. A. M., 79–80
Dirichlet-type boundary conditions, 12, 44, 85
Dispersion relation, 70
Dry mode analysis, 205–206

221

interace of two fluids, or
fluids/solid surface, 15
interfacial boundary conditions,
20
mean free surface, 16–17
overview, 14–15
two-layer fluids, 67–68
velocity of water particles, 16, 17
wave interaction with rigid
structures, 43
Wave breaking. *see* Breaking, waves
Wave classification, 19
Wave equations, 12–13
one-dimensional, 13
Wave numbers, 75, 108, 181
Wave structures
breakwaters; *see* Breakwaters
floating structures, contact with;
see Floating structures
fluids, contact with, 1
interactions; *see* Interactions,
wave structures
linearised water wave theory; *see*
Water wave theory,
linearised
ocean surface waves; *see* Surface
waves
scattering; *see* Scattering
Wave transformations, 19
breaking; *see* Breaking, wave
diffraction; *see* Diffraction, wave
physical properties of, 22
scattering; *see* Scattering, wave
Wave trapping. *See* Trapping, wave
Wave-ice interactions
Arctic and Antarctic regions, 3
flexural gravity waves created
by, 150, 151
Waves, gravity. *See* Flexural gravity
waves; Surface gravity
waves
Wide spacing approximation, of
barriers, 141–144, 171
Wiener-Hopf technique, 3, 4, 13

W
Young's modulus, 23, 31, 150

X
Zero normal velocity, 94, 158